SpringerBriefs in Applied Sciences and Technology

PoliMI SpringerBriefs

SpringerBriefs present concise summaries of cutting-edge research and practical applications across a wide spectrum of fields. Featuring compact volumes of 50 to 125 pages, the series covers a range of content from professional to academic.

Typical publications can be:

- A timely report of state-of-the art methods
- An introduction to or a manual for the application of mathematical or computer techniques
- A bridge between new research results, as published in journal articles
- A snapshot of a hot or emerging topic
- An in-depth case study
- A presentation of core concepts that students must understand in order to make independent contributions

SpringerBriefs are characterized by fast, global electronic dissemination, standard publishing contracts, standardized manuscript preparation and formatting guidelines, and expedited production schedules.

On the one hand, **SpringerBriefs in Applied Sciences and Technology** are devoted to the publication of fundamentals and applications within the different classical engineering disciplines as well as in interdisciplinary fields that recently emerged between these areas. On the other hand, as the boundary separating fundamental research and applied technology is more and more dissolving, this series is particularly open to trans-disciplinary topics between fundamental science and engineering.

Indexed by EI-Compendex, SCOPUS and Springerlink.

Springer, in cooperation with Politecnico di Milano, publishes the PoliMI SpringerBriefs, concise summaries of cutting-edge research and practical applications across a wide spectrum of fields. Featuring compact volumes of 50 to 125 (150 as a maximum) pages, the series covers a range of contents from professional to academic in the following research areas carried out at Politecnico:

- Aerospace Engineering
- Bioengineering
- Electrical Engineering
- Energy and Nuclear Science and Technology
- Environmental and Infrastructure Engineering
- Industrial Chemistry and Chemical Engineering
- Information Technology
- Management, Economics and Industrial Engineering
- Materials Engineering
- Mathematical Models and Methods in Engineering
- Mechanical Engineering
- Structural Seismic and Geotechnical Engineering
- Built Environment and Construction Engineering
- Physics
- Design and Technologies
- Urban Planning, Design, and Policy

Francesco Vergani

Reframing Urban Ecologies

Plant Agency and More-than-Human Perspectives in Co-designing Public Spaces

Springer

Francesco Vergani
Department of Design
Politecnico di Milano
Milano, Italy

ISSN 2191-530X ISSN 2191-5318 (electronic)
SpringerBriefs in Applied Sciences and Technology
ISSN 2282-2577 ISSN 2282-2585 (electronic)
PoliMI SpringerBriefs
ISBN 978-3-032-18923-3 ISBN 978-3-032-18924-0 (eBook)
https://doi.org/10.1007/978-3-032-18924-0

This Springer imprint is published by the registered company Springer Nature Switzerland AG
The registered company address is: Gewerbestrasse 11, 6330 Cham, Switzerland

Foreword

It is with great intellectual satisfaction that I introduce *Reframing Urban Ecologies: Plant Agency and More-Than-Human Perspectives in Co-designing Public Spaces*. This book, authored by a former doctoral student whose research trajectory I have had the privilege to supervise within the Polimi DESIS Lab at the Politecnico di Milano, represents a mature and rigorously articulated contribution to ongoing debates at the intersection of design theory, environmental humanities, and urban ecology. From the outset of his doctoral work, Francesco demonstrated a capacity to engage critically with entrenched assumptions in design discourse and to situate his inquiry within a wider constellation of theoretical and methodological challenges. The result is a volume that not only synthesises diverse strands of scholarship but also advances an original framework for rethinking the role of nonhuman life, particularly plants, in processes of urban cohabitation and co-design.

The book's point of departure is a fundamental reconsideration of plants as dynamic, adaptive, and relational organisms. The author convincingly shows how plants continuously act upon and within their environments, displaying forms of agency that, although often overlooked, are both ecologically consequential and theoretically provocative. The urban landscape provides a particularly fertile context for observing these interactions, revealing plants as beings that resist, adapt, and transform despite the stressors characteristic of contemporary cities. By foregrounding these dynamics, Francesco challenges the persistent reduction of plants to aesthetic or infrastructural elements, arguing instead for their recognition as active co-constituents of urban public space.

Simultaneously, the work engages with a growing body of scholarship seeking to move beyond the epistemological and ontological constraints of human-centred design. Drawing on debates surrounding the Anthropocene, ontological design, and post-human and decolonial perspectives, the book maps the conceptual terrain within which more-than-human and multispecies design have emerged as critical frameworks.

This work will undoubtedly be of interest to scholars and practitioners concerned with environmental design, urban planning, critical plant studies, and multispecies approaches. Its strength lies not only in its conceptual sophistication but also in its capacity to integrate scientific, philosophical, and design-based forms of knowledge without subsuming one under another. In doing so, it opens new possibilities for understanding how design might contribute to more equitable and ecologically attuned urban futures. Within the context of the Polimi DESIS Lab, these reflections have also played a meaningful role in pushing the boundaries of our collective research, opening new frontiers of observation and inquiry into the relational dynamics between humans, plants, and the urban environments we co-inhabit.

Department of Design
Politecnico di Milano,
Milano, Italy

Davide Fassi

Preface

The role attributed to plants within urban environments has long been a subject of debate and contention. Should they be understood as vital contributors to human well-being, or merely as aesthetic features that enhance the attractiveness and liveability of cities? More fundamentally, should they be recognised as living organisms in their own right, or simply as planning *assets* that enhance citizens' physical and mental well-being? Unlike other structural features of urban public spaces, such as architecture, street furniture, or lighting, which generally remain fixed, plants are dynamic. They change and adapt gradually in response to seasonal cycles, environmental conditions, and external stimuli. They are always present, yet always in transformation: sometimes strong and flourishing, at other times fragile and unstable. Upon being situated in a specific location, they strive to adapt, demonstrating resilience in the face of frequently challenging conditions. In short, plants do not merely exist in the world: they *inhabit* it. They explore the soil in search of resources and structural anchorage, absorb water and sunlight to nourish themselves, and modify their surroundings by releasing organic compounds into the atmosphere. Plants are constantly engaged with their environment through a complex network of actions, behaviours, and relationships. Although this may appear distant from a human perspective, plant life is far from silent, passive, or inert, as they seem to act with intention and purpose.

As humans, we recognise the vital role plants play in our survival; however, it is challenging to perceive them as active participants in our daily existence. Our tendency to underestimate nonhuman entities frequently results in the perception of plants as mere ornamental features, rather than as living organisms possessing their own agency. And yet this active capacity of plants is visible all around us in urban public spaces: tree roots breaking through pavements in search of nutrients, branches stretching towards the sun, and flowers drawing in pollinators. These are just a few of the many behaviours that reveal their agency. To what extent, then, does our anthropocentric worldview prevent us from recognising an active—and in many ways *design-like*—capacity not only in plants, but across all forms of life?

In recent decades, the fields of design and urban planning have developed a growing sensitivity to environmental issues, opening both epistemic and practical spaces for reimagining how humanity might contribute to shaping a world that is biologically richer and more resilient. Contemporary design practices and biodiversity strategies are increasingly embedded within urban contexts and municipal planning guidelines, promoting ecological sustainability and reintroducing plants in relation to their entanglements with other nonhuman life forms. Yet, while these practices help strengthen urban biodiversity, they remain primarily framed by anthropocentric reasoning, valuing nonhuman life mainly for its instrumental benefits to human physical and psychological well-being. As a result, they leave little room to engage with emerging debates on *vegetal* intelligence and agency. Plants, in particular, are often constrained or suppressed by the toxic conditions of urban environments, where continuous stress factors affect not only them but also other species (humans included).

Although such considerations are still marginal within most administrative and political agendas—with only a few notable exceptions—participatory design processes for urban public spaces may provide a practical framework for fostering greater sensitivity among local communities to more-than-human perspectives. By foregrounding the active capacities of plants and other life forms, these processes might potentially position them as co-agents in the co-envisioning and co-design of future scenarios of coexistence. In doing so, they could disrupt anthropocentric paradigms and contribute to the advancement of more equitable, multispecies futures.

In recent years, my research has increasingly turned towards exploring plant agency from multiple perspectives. Although this was not the central focus of my research at the outset, I gradually came to approach it with a sense of urgency. I began to reflect on how to effectively draw on my experience as a member of the Polimi DESIS Lab research team (Department of Design, Politecnico di Milano) and how to situate this inquiry within the broader field of design research. Having worked primarily on Design for Social Innovation projects—initiatives aimed at supporting situated communities to act proactively, often through participatory approaches—I shifted my attention to how nonhuman agents are acknowledged within participatory design processes for urban public spaces. My decision to focus on the *vegetal* world, first in my doctoral research and now in ongoing research projects, has been motivated by both personal concerns and the critical role that plants and other nonhumans, such as animals and bacteria, play within the broader ecosystem.

Over the past decade, more-than-human design has emerged as a conceptual lens that weaves together principles from planet-centred, life-centred, and other emergent approaches, offering a coherent framework for rethinking design's role in shaping multispecies and planetary futures by acknowledging other agents as active collaborators within design processes. However, within deeply entrenched anthropocentric paradigms, this remains particularly difficult to put into practice. Introducing more-than-human perspectives into scientific research inevitably entails additional tensions and limitations. The challenge is compounded by limited understanding of species-specific behaviours, insufficient attunement to nonhuman forms of life, and the persistent marginalisation of nonhuman agency.

Most notably, there is a clear friction between the dominant tradition of Western, empirical knowledge and emergent perspectives—unfamiliar to many Western scholars—that stem from Indigenous knowledge. As a European researcher, I have primarily approached the themes of this book from a scientific perspective. However, my work occasionally embraces forms of knowledge that are less easily verifiable yet still profoundly relevant to the issues at hand. This careful negotiation of personal standpoint reflects the very nature of the themes under investigation. Themes that are, by definition, in a state of ongoing exploration and often situated on uncertain epistemic ground. They draw on bodies of knowledge that are not yet firmly established, even within their own disciplinary domains, with plant neurobiology serving as a particularly salient example.

The book is based on strong scientific foundations, but it frequently draws on the creative and speculative power of design to envision what might happen in the future. Concepts such as human–nonhuman interaction, species relationships, and the status of sentient beings are rooted in an ambiguous and still largely underexplored field. Bridging this gap is one of the most challenging tasks for designers, but it is also a crucial step towards embracing a truly more-than-human perspective.

Within this framework, *Reframing Urban Ecologies: Plant Agency and More-Than-Human Perspectives in Co-designing Public Spaces* presents an investigation of this approach, with a specific focus on plants and humans as primary agents. The book underscores the importance of recognising and cultivating interspecies relations as a foundation for designing plural, multispecies urban environments. Although it engages with the domains of more-than-human and multispecies design, its purpose is not to present case studies. Instead, it situates itself between the dissemination of knowledge to enhance the recognition of plants as active participants in design and the integration and evaluation of that knowledge within established practices of urban communities.

In this sense, the study highlights the ambiguous role of plants in and for urban ecosystems and their stakeholders, offering a lens through which to reimagine these ecologies by placing relationality at the very centre. The *Relational Atlas* introduced here serves as a framework for decentring the human and unsettling power dynamics in urban design by reframing the design space as a *relational space*: a moment in which the shared interests of human and nonhuman agents can converge at a specific spatial and temporal node. This theoretical framework aims to inform participatory design practices within multispecies communities, challenging anthropocentric perspectives and opening pathways towards new forms of urban ecology. Through the construction of this *Atlas*, the book seeks to nurture a discourse that enables humans and designers to recognise points of connection across species while resisting the inner impulse to place themselves at the centre. The book is organised into six chapters.

Chapter 1 lays the theoretical foundation of the book, drawing on debates in the humanities and social sciences to challenge anthropocentric and utilitarian views of design. It introduces more-than-human and multispecies design as approaches that expand the scope of concern beyond the human, recognising the agency, needs, and perspectives of nonhumans.

Chapter 2 shifts to the natural sciences, synthesising contemporary literature on plant biology. It reframes plants not as passive or aesthetic features, but as active participants in ecological systems, highlighting their agency by bridging scientific knowledge with speculative interpretation.

Chapter 3 presents the *Relational Atlas*, a framework that translates plant biology into a complex network of relations. By outlining plant *strategies*, it positions plants as relational agents and provides ways for integrating their perspectives into design and planning processes.

Chapter 4 turns explicitly to urban contexts, examining how planning has historically engaged (or failed to engage) with nonhuman life. Through an analysis of design approaches at different scales and ethnographic observations, it identifies the *stressors* that constrain plants in cities.

Chapter 5 examines how participatory design can promote *middle-out* engagement, connecting grassroots practices with administrative policies. It introduces a co-design activity aimed at raising human awareness of multispecies communities, with particular emphasis on recognising plants as stakeholders in the urban redesign of a small park.

To conclude, Chapter 6 revisits more-than-human and multispecies design in light of the book's findings, offering reflections on these new approaches to support interspecies relations and underscoring the need for design to move towards more equitable, regenerative, and multispecies futures.

Milano, Italy Francesco Vergani

Acknowledgements

I owe much of the inspiration for this volume to the multitude of life forms that have populated my experiences in the cities where I have lived over the years. From trees that, with quiet strength and resilience, bend and reshape streets and pavements, to the small plants that stubbornly carve out space within the folds of urban structures, these presences have gradually redefined the way I see the world. Not so long ago, I would never have imagined embarking on a research journey in this particular field, nor doing so by adopting—as far as possible within the limits of my *being human*—this perspective. If this has been possible today, I owe it entirely to them.

I also wish to express my deepest and most sincere gratitude to all the humans who have accompanied me along this path. First and foremost, Davide Fassi, for allowing and encouraging me to embark on this journey long before I could imagine it myself and for remaining a constant, attentive, receptive, and open presence. I thank Virginia Tassinari for opening the doors to a realm of wonder and curiosity, offering continuous support marked by a generosity of spirit, intellectual sharpness, and wit. I thank Elisa Giaccardi for her dedication, kindness, contagious passion, and the constant encouragement that have accompanied the most recent years of this journey.

My heartfelt thanks go to the entire research group of the Polimi DESIS Lab and to my colleagues in the Department of Design, who have walked alongside me over the years, contributing not only to my professional growth but also to making each working day an experience rich in exchange and joy: Vanessa Monna, Ambra Borin, Laura Galluzzo, Annalinda De Rosa, Claudia Mastrantoni, Valentina Ferreri, Valentina Facoetti, Maria Maramotti, Irene Bassi, Nadia Pirovano, Mathew Spialtini, Andrea Manciaracina, and all the others—so many—who have shared ideas, time, and energy.

I also thank my colleagues in the DESIS Network—Louise St. Pierre, Francesca Valsecchi, Tokushu Inamura, and Ezio Manzini—for the intense, stimulating, and illuminating conversations, often late at night, which helped to expand and refine the conceptual framework within which this volume is situated.

Special thanks go to Aitor Acilu for allowing me, with great openness and trust, to test certain ideas within a different university context, and to Elisa Bertolotti for welcoming me to the beautiful island of Madeira, creating the conditions for the emergence of new perspectives and reflections.

I am grateful to the community of the Nolo neighbourhood in Milan and, in particular, to Desiré Musumeci, Franco Beccari, James Conrad, and Maria Grazia Allegra, for their active participation, generosity, and the invaluable support they have offered over the years.

I would also like to thank most of the students with whom I have had the pleasure of working. Their willingness to engage, their openness to new horizons, and their unshakeable sense of hope in what they may spark in the future have deeply supported and comforted me at the moment when I began—with apprehension, yet also with a strong sense of responsibility—to put into practice what I had observed, nurtured, gathered, and developed.

Finally, I extend my gratitude to all the academics and professionals who have taken part in this process, enriching the work with unique and invaluable contributions, dialogues, and insights.

Milano, Italy
Francesco Vergani
November 2025

Competing Interests The author has no competing interests to declare that are relevant to the content of this manuscript.

Contents

Chapter 1
Challenging Anthropocentrism in Design

Abstract This introductory chapter establishes the theoretical foundation of the book, situating design within broader debates in the humanities and social sciences. It critically examines the anthropocentric and utilitarian perspectives that have long dominated design discourse, demonstrating how such views limit our capacity to address today's complex ecological and social challenges. Drawing on insights from philosophy, anthropology, and design studies, the chapter introduces the fields of more-than-human and multispecies design, articulating their efforts to move beyond a human-centric understanding of the world. These approaches call for a radical rethinking of design's purpose, extending its focus beyond human needs towards the recognition of the agency, needs, and perspectives of nonhumans. By foregrounding interdependence and relationality, the chapter argues that design must evolve into a practice that embraces multispecies coexistence and planetary well-being. At the same time, it provides a critical lens for engaging with the subsequent chapters of the book, while also acknowledging the limits and constraints of a more-than-human perspective.

1.1 Introduction. The Evolving Meaning of Design

Design encompasses a wide range of meanings and applications, reflecting the diverse and creative ways it is interpreted across fields and contexts. In Italian, the verb *progettare*—from the Latin word *projèctus*, the "action of throwing ahead"—emphasises an intention conceived in the present and projected into the future. The term *design* entered the English language in the early sixteenth century, where, among its various meanings, it was used to refer to a plan, intention, purpose, or goal. It has since become closely associated with creating plans for the development of objects and systems, a process traditionally carried out by humans. As such, design is most often discussed in relation to human activities and creations [1–3]. The Second Industrial Revolution marked the emergence of design as a practice aimed at satisfying human needs and aspirations, first through technology-driven approaches, and later through human-centred practices [4].

F. Vergani, *Reframing Urban Ecologies*, SpringerBriefs in Applied Sciences and Technology, https://doi.org/10.1007/978-3-032-18924-0_1

Today, the term encompasses multiple meanings difficult to comprehend entirely, especially as it is also broadly defined as any systematic or intentional arrangement of elements or features, whether influenced by humans or other living organisms. Drawing on anthropologist Arturo Escobar's definition [5], "design is about creating cultural meanings and practices," which inherently links it to the concept of culture. The word *culture* derives from the Latin *colere*, originally used to indicate the act of "cultivate the soil" or "care for plants," and only later expanded to refer more generally to "the set of intellectual knowledge that a person has acquired, reworked [...] into a constitutive element of their moral personality, spirituality, and aesthetic taste." As Ramsey [6] defines it, culture is "[the] information transmitted between individuals or groups, where this information flows through and brings about the reproduction of [...] the behavioural trait." That said, across the centuries, both *culture* and *design* have mainly been regarded as virtues of the human intellect.

While this definition transcends symbolic and cognitive dimensions, its dominant interpretation within the Western tradition has historically reinforced an anthropocentric bias, framing both design and culture as exclusively human endeavours. As Roudavski [7] observes, anthropologists such as Descola [8] and Ingold [9] have long highlighted that many societies—particularly those informed by Indigenous worldviews and animist traditions [10]—do not draw such rigid boundaries. From these perspectives, nonhuman beings such as animals, plants or rocks are acknowledged as co-constitutive of culture and as active participants in shaping worlds. The suppression of such views within Western epistemologies has perpetuated the illusion that only humans design and that culture is solely a human domain. This anthropocentric perspective has been repeatedly challenged in design theory: Fry's [11] notion of *defuturing* exposes how human-centred design contributes to planetary degradation. At the same time, Escobar [5] calls for autonomous and ontological design that learns from and with nonhumans. Likewise, Haraway's [12] speculative *fabulation* and Tsing's [13] practice of *noticing* reframe design as a situated, multispecies endeavour.

In today's fragile and rapidly shifting planetary context, marked by intensifying ecological crises and the urgent need to rethink coexistence, the roles of design and culture call for critical reconsideration and gradual redefinition. Their scope can no longer exclude other living and non-living beings, whose agencies and contributions are indispensable to the flourishing of multispecies and planetary futures.

This first chapter is grounded in the idea that design can, and should, be open to broader interpretations. Moving beyond a strictly human-centred perspective, design should be framed not just as a solutionism-based practice, but as a relational process in which various entities actively shape the world, both influencing and being influenced by others. Intertwined with insights from the humanities (philosophy) and the social sciences (anthropology, political science, and sociology), the chapter addresses the pressing environmental, political, and social challenges of our time, exploring how design can, and must, engage with these issues while critically examining its responsibilities in pursuing these goals [14].

1.2 Design in the Age of the Anthropocene

From a Western perspective, design is closely tied to human cognitive and creative skills in solving problems and creating new products, environments, and experiences [15, 16]. Humans are seen as the *governing force* over their surroundings, shaping relationships within their own species while "gradually [rising] above, and [bringing] under control, both their own nature, in the process of civilisation, and the nature around them, in the domestication of animals and plants" [9]. This notion of human predominance—accelerated by technological advancements starting from the Second Industrial Revolution [17]—has led to a profound imbalance among living species, with significant impacts on Earth's geological conditions [18]. In just a relatively short period, human activity has drastically altered ecological processes [19, 20], driving extinction rates of *wild* species to levels between one hundred and one thousand times higher than in the pre-industrial era [21].

This destructive human impact on Earth has, for decades, been encapsulated in the concept of the Anthropocene, a term coined by Crutzen and Stoermer [22] to describe a proposed geological epoch beginning with significant human influence on the planet's geology and ecosystems, including but not limited to climate change [23, 24]. The ecological disruptions already visible today, most notably those driven by global warming, are expected to leave stratigraphic traces that will endure for millions of years. As a direct and indirect consequence of human-induced climate change [25], the biosphere is undergoing massive, irreversible transformations [26], jeopardising the survival and well-being of humans as well as countless other life forms [27].

For much of the scientific community, the Anthropocene is primarily attributed to industrial and technological expansion, which has released vast quantities of carbon dioxide and other greenhouse gases into the atmosphere [25]. Others, however, such as McManus [28], trace its origins to the drastic drop in atmospheric carbon dioxide following the large-scale genocide of Indigenous peoples in the Americas during the seventeenth century by European colonisers, which also triggered an ecocide of local ecosystems. In either case, the Anthropocene can be viewed as an unintended and profoundly irrational outcome of human activity [29], whether it emerges from destructive sociocultural processes, such as colonialism, or from the relentless pace of technological growth. However, in 2024, the International Commission on Stratigraphy and the International Union of Geological Sciences voted against officially renaming the current geological epoch (the Holocene) due to a lack of clarity and consensus over the formal definition of the Anthropocene and its proposed start date. Nevertheless, despite this formal rejection, the term Anthropocene remains widely used in both scientific and cultural discourse to describe humanity's impact on the planet and its influence on Earth's ecosystems.

The environmental crisis underscored by the concept of the Anthropocene—whose visibility has grown globally through movements like *Fridays for Future* and, more recently, in the wake of the COVID-19 pandemic [7, 30]—has intensified the call to reassess humanity's role on the planet. This paradigm shift is not entirely new, but the accelerating pace of climate change compels us to urgently explore alternative approaches that might help slow the trajectory toward potentially catastrophic futures for our species. This shift calls for a fundamental rethinking of how we engage with, and therefore design, the environments, embracing alternative worldviews and fostering new ways of reframing collaboration with other life forms [7, 31–35].

1.3 A Matter of Perspective

As noted in the introduction, to *design*, in its broadest sense, can be understood as *acting to create something for a future purpose*, an action marked by the deliberate intentionality of a person performing a gesture to achieve a specific goal. However, its meaning has shifted over time, shaped by the conventions and traditions of the cultural groups in which it is practised. This variability extends to the diverse actions and practices that design encompasses, with expressions differing across contexts.

In recent years, design has increasingly engaged with questions raised by disciplines such as anthropology and philosophy, emphasising the need to move beyond a strictly Western perspective [5, 36, 37]. There is a growing urgency to *decolonise* both our ways of acting and our approaches to design, bringing forward the uses and practices from contexts that are too often overlooked or suppressed within their own communities [5]. *Decolonising* design involves "problematizing the affective and conceptual operations that form the basis of our relations with the world and questioning the essentialist or instrumentalist approaches that have been naturalised previously" [38].

From this perspective, the concept of the *Pluriverse* [5, 29, 36, 39] emerges as the coexistence of multiple, diverse design practices and lesser-known design traditions that reject any singular, closed definition. This vision departs from a purely utilitarian or aesthetic paradigm (typical of "patriarchal Western capitalism" [5], particularly evident in the modernity project [40]) and instead aligns with meaningful, participatory, and situated experiences [41]. These approaches are often connected with *indigenous knowledge* [39, 42] and *matristic cultures* [43], rooted in inclusive, relational methods that reframe the connection between humans and their surroundings, both living and non-living. They reject the subject/object dichotomy embedded in Western-centric, anthropocentric thought [11], and instead acknowledge a complex system of interdependent worlds. In line with Escobar, French philosopher and anthropologist Bruno Latour [40] also called for a new perspective, introducing the figure of the *terrestrial* as a political actor aware of contemporary challenges, recognising itself as part of a vast, living entity. Latour framed the *terrestrial* as an alternative force, countering the rigid opposition between the *Global* (the modernisation project driven by globalisation) and the *Local*, which is linked to soil and bottom-up practices, but is free from the national boundaries and identities of the past. From his perspective, the term *terrestrial* is broad, encompassing not only humans but also animals, plants, and the concept of *Gaia*, acknowledging the interconnectedness of all life forms in sustaining planetary well-being. The theory of *Gaia* (the Mother Earth goddess) proposes that living organisms interact with their inorganic surroundings on Earth to form a synergistic and self-regulating complex system that helps to maintain and perpetuate the conditions for life on the planet [44].

What connects many of these theories and practices discussed is the recognition that interactions are not exclusively human-driven but are often initiated by other entities. These may be individual living beings, such as animals or plants, or larger, more complex entities, such as rivers. In each case, an external action that affects humans is acknowledged. To expand design beyond an exclusively human practice, it is therefore essential to consider how the surrounding environment not only influences us but also shapes the very act of design itself. In this regard, "ontological design" [5, 33, 45] plays a pivotal role in framing the relationship between human beings and their lifeworlds. Willis [45], and later Escobar [5] identify in this ontological turn a dual movement of design: a fluid exchange in which we (as human beings) design our world, while our world acts back on us and designs us. Here, design extends beyond the creation of objects, systems, or environments for human purposes: it is about shaping the very conditions of existence and, in turn, being shaped by them. The notion of the environment as a *design force* also underpins systemic and holistic approaches to design, which address complexity by tracing relationships among the interdependent components of a system, viewed as an inseparable whole.

1.4 More-Than-Human Design

The growing urgency to reframe the role of design in response to the multiple, interwoven ontologies calls for approaches that move beyond the limitations of human-centric paradigms, which are increasingly recognised as inadequate [46–49]. Tironi and Garretón [50], for example, advocate a shift toward "planet-centred design", which foregrounds the interdependencies among humans, nonhumans, technologies, and ecosystems, actively challenging the anthropocentric assumptions embedded in conventional design practice. Similarly, Borthwick et al. [51] and Lutz [52] advocate for "life-centred design," emphasising long-term, systemic thinking and inclusivity across species and communities to incorporate nonhumans and *safeguard* planetary well-being. Over the past decade, such perspectives have contributed to a significant reorientation of design discourse and practice toward more-than-human considerations. Within this broader frame, more-than-human design [32, 53, 54] offers an overarching conceptual lens to integrate the principles of planet-centred, life-centred, and other emerging approaches, providing a coherent foundation for rethinking design's role in fostering multispecies and planetary futures.

More-than-human design is an approach that shifts the focus of design beyond exclusively human concerns to incorporate the perspectives, needs, and agency of nonhuman actors, including both biological entities (such as animals, plants, and ecosystems) and synthetic components (such as materials and technologies). Looking at the world through a more-than-human lens implies challenging the long-standing dualistic paradigm that positions the *natural* and the *artificial* as distinct and opposing realms [55–58], a conceptual division deeply embedded in dominant cultural and design discourses. From this perspective, philosopher Emanuele Coccia [59] argues that "there is no natural environment [as the] world is always, in every part, conceived, designed, and built". He views the world as an artificial (meaning *continuously modified*) product, shaped by an ongoing metamorphosis in which not only humans, but all materials and forms on the planet, are active protagonists [59].

In line with the principles of *ontological design* and as formalised in *actor-network theory* [60, 61], the relationship between any subject and its surroundings is constant, as "space is always designed and built by other species, and for other species than the one that occupies it" [59]. This view shifts the understanding of the world from a dichotomy between the human (*artificial*) and the nonhuman (*natural*) toward seeing it as a collective of things [8, 32, 62], an *assemblage* defined as a fluid, open-ended network of vibrant materials, both human and nonhuman [60, 63–66]. Within this framework, humans emerge as just one factor in a larger system of relationships and interactions [67], one of many actors that act and organise together (and therefore design), sharing the delegation of power and agency as understood by the aforementioned *actor-network theory* [32].

Design is only now gradually opening up to new actors and networks that have yet to be fully recognised as legitimate participants in the design process [4]. In this regard, the term *agent*—derived from *agency*, which is intended to signify the capacity to act as a relational, distributed, and more-than-human achievement [68]—can serve to acknowledge the active role of nonhumans that are typically overlooked. While design, particularly within Human-Computer Interaction, has long engaged with synthetic agents, such as emerging technologies and artificial entities that challenge traditional human–technology boundaries [69–73], the inclusion of biological nonhumans is a far more recent development. Within this context, design is beginning to embrace a multispecies perspective [7, 74–77], exploring new ways of relating to agents across fundamentally distinct species. Yet, from this perspective, it is essential to examine the dynamics of "interspecies relationships" [7, 8, 78, 79] that occur among its diverse constituents. This "multispecies affair" [59] shows how diverse lifeforms are bound together in what Escobar [80] describes as a continual state of connection, a "flux of changing forms, forces, behaviours [...] always in the process of dependent co-arising". Such interdependence also manifests as "coexistence" [81–84] between agents: a demanding proposition that is difficult to realise from a non-anthropocentric standpoint, as coexistence requires ongoing negotiation, balancing, and diplomacy. We must remain aware of our role, rethinking how we, as humans, relate not only to one another but also to all other life forms.

This ontological turn, as discussed earlier, should be understood first and foremost as a process of recognising the implications of being *terrestrial* within a broader *assemblage*. Maria Puig de la Bellacasa [62], building on Joan Tronto [85], frames this responsibility through the concept of *care*, defined as "everything that we do to maintain, continue, and repair [the] world so that we can live in it as well as possible" [62]. Care is interwoven into a complex, life-sustaining web, encompassing all actions and practices aimed at addressing the environmental crises of the Anthropocene and fostering the radical interdependence between agents. This calls for a process of "re-earthing" [80], rejecting the anthropocentric Western worldview [86] and acting ethically toward all agents [87]. As Ingold [88] observes, "we inhabit our environment: we are part of it," and must therefore act from an ethical standpoint, also considering the needs and rights of nonhumans [67]. From this perspective, we are urged to cultivate what Jain [89] terms a "multispecies politics of care", an ethical and political orientation that extends the scope of concern, responsibility, and reciprocity beyond the human. Such an approach calls for a reconfiguration of traditional political discourse to encompass "multispecies justice" [90], recognising the rights, needs, and vulnerabilities of diverse species within shared socio-ecological systems. It also requires attentiveness to "ecological entanglements" [91], acknowledging the complex webs of interdependence that bind living and non-living entities, as well as a deeper understanding of human–environment interdependencies [92] that situate human flourishing within broader ecological processes. Central to this shift is the recognition of the *agency* of nonhumans [49], whose capacities to act and to influence outcomes challenge anthropocentric assumptions and invite more inclusive, relational, and ecologically attuned forms of design.

1.5 Designing *for* or Designing *with*?

Designing through a more-than-human lens means recognising other agents as active collaborators in the design process. Yet, from an anthropocentric perspective, this proves particularly difficult to implement due to a wide range of challenges and limitations. While the idea underpinning more-than-human design is now theoretically articulated, in practice it remains ambiguous and complex to realise [4, 7, 74, 79, 93–96]. Within this context, two main approaches—among several others—can be identified in design practices aligned with the more-than-human paradigm: *designing for* and *designing with*.

Designing for refers to an approach in which designers position themselves as the leading actors in the process, taking on the responsibility of planning for other nonhuman agents. In this framework, plants, animals, or ecosystems are regarded as recipients of the design, often interpreted through scientific or technical knowledge that mediates their needs and behaviours. Data collected through research helps support nonhuman agencies, providing a body of knowledge that is useful for understanding their requirements and objectives. Although this represents progress beyond a purely anthropocentric approach, it still maintains an asymmetry of power, as designers and human communities remain the primary decision-makers. At the same time, nonhumans are primarily positioned as objects of care or protection.

Designing with, on the other hand, implies co-agency: nonhumans are acknowledged as partners in the design process, carrying needs, constraints, and possibilities that actively shape the project. This shifts the paradigm towards a relationship of interaction, where design is not only carried out *for* others but *with* others. In this approach, design becomes a form of attunement to the rhythms, signals, strategies, and behaviours of the nonhuman. Here, the nonhuman emerges as a co-designer (albeit often indirectly) since its modes of life, adaptation, and communication directly inform the project's outcomes. On a biological level, and drawing on ideas developed in connection with the concept of ontological design, *designing with* takes shape as a "multispecies interaction" [97], an approach that seeks to emphasise sensitivity to the diverse ways in which living beings connect and influence one another. While more-than-human design leans towards the *with* rather than the *for*, fully understanding the dynamics of *with* remains particularly challenging, mainly due to factors that are, in many ways, inherent to the subject itself. More than in other branches of design, designing *with* nonhuman agents is a demanding challenge [98], fraught with ambiguities and critical issues. To this day, many of these complexities are hard to overcome.

1.6 Questioning the Post-Anthropocentric Turn: Complexities and Limitations

As highlighted in this chapter, it is crucial to recognise that a post-anthropocentric understanding of design exists at the threshold between Empirical-rooted scientific knowledge and indigenous ways of knowing. This overlap inevitably generates tensions on both sides, especially within Western scientific traditions, since much of the knowledge underpinning nonhuman agency remains fragile, provisional, and in constant flux within its disciplinary domains. Expanding design through a more-than-human lens requires facing a series of persistent limitations that continue to shape and constrain this emerging field. Since these challenges and tensions recur throughout the volume, it is helpful to outline them explicitly at the outset.

First, engaging with design inevitably entails addressing questions of, qualities that are often recognised as belonging solely to the human species. This limitation, clearly articulated in the notion of "anthropomorphism" [99], is embedded in our anthropocentric stance, which places human traits and design abilities at the centre of everything. This book, therefore, draws on the idea of design as described by Westerlaken [77] (building on Abram [100]) as an intentional process of "negotiations, exchanges, articulations, and nourishments", a perspective that resonates with the concept of *multispecies worlding*. In this sense, design becomes a practice of co-creating and sustaining worlds shared by multiple species, grounded in reciprocal relationships and in the continual reconfiguration of life's interdependencies. However, to address the topics of this volume, we must move beyond anthropocentric conceptions, acknowledging that humans cannot be the only beings endowed with such qualities. This bias may represent a biological constraint that is presumably common to all species, given that each is fundamentally focused on preserving its own kind. Every living species strives for its own well-being, most often relying on relationships with other forms of life to support this aim. Recognising this limitation can help us understand that the life trajectories of different species are not so dissimilar as they may seem, but rather manifest in situated, unique, and distinctive ways.

Moving towards more-than-human considerations means engaging with the interactions between agents and nonhuman entities. These may be biological (belonging to other species such as plants and animals) or synthetic (such as Artificial Intelligence). As discussed earlier, our knowledge of how these different agents interact is often fragmented. For this reason, when addressing more-than-human design, discourse frequently drifts into speculation and fabulation [12, 101, 102]. Currently, establishing a shared medium of interaction between species remains highly challenging. When we speak of interaction in human terms, we usually refer to a form of codified language—whether audio, physical, or tactile—which is nearly impossible to employ across interspecies relationships. The inability to clearly understand the needs and trajectories of other species is a fundamental limitation within the more-than-human approach. As humans, we often rely on scientific research to understand the well-being of different species; yet, it remains impossible to fully take the perspective of those agents. This work of *translation* and *mediation,* therefore, emerges as a significant challenge within such relationships, one that that must be carefully acknowledged and addressed.

Ultimately, discussing *design* requires addressing the (typically human) power dynamics that shape these practices. Relationships between species are almost always unbalanced, with human dominance often prevailing. Beyond anthropocentrism, this issue is closely tied to the absence of human policies and legal frameworks that recognise and safeguard the rights of nonhumans, thus preventing their inclusion in participatory processes that could mediate between human and nonhuman needs to enable collaborative and mutualistic outcomes. As Huffman observed nearly 30 years ago, [103]: "unless we develop the capacity to communicate with these other forms of life, we will not be able to accommodate nonhuman rights holders in our system of rights because they will not have the essential capacity to appeal for enforcement of violated rights." However, the contribution of contemporary more-than-human design lies in its ability to open up spaces for discussion and design practices that can foster regenerative forms of coexistence. And this is precisely the aim of this book: to reframe and imagine new urban ecologies through conversations and practices centred on plants and their behaviours.

References

1. Brown T (2008) Design thinking. Harv Bus Rev 86:84
2. Heskett J (2005) Design: a very short introduction. Oxford University Press
3. Nelson HG, Stolterman E (2014) The design way. Intentional change in an unpredictable world, 2nd edn. The MIT Press, Cambridge, MA
4. Veselova E, Gaziulusoy İ (2021) Bioinclusive collaborative and participatory design: a conceptual framework and a research agenda. Des Cult 14:149–183
5. Escobar A (2018) Designs for the pluriverse: radical interdependence, autonomy, and the making of worlds. Duke University Press
6. Ramsey G (2013) Culture in humans and other animals. Biol Philos 28:457–479
7. Roudavski S (2020) Multispecies cohabitation and future design. In Boess S, Cheung M, Cain R (eds) Synergy - DRS International Conference 2020. https://doi.org/10.21606/drs.2020.402
8. Descola P (2013) The ecology of others. Prickly Paradigm Press
9. Ingold T (2000) The perception of the environment. Essays on livelihood, dwelling and skill. Routledge, London
10. Harvey G (2014) The handbook of contemporary animism. Routledge
11. Fry T (2020) Defuturing. A new design philosophy. Bloomsbury, London
12. Haraway D (2013) SF: science fiction, speculative fabulation, string figures, so far. Ada J Gender New Media Technol 3:18
13. Tsing AL (2015) The mushroom at the end of the world: on the possibility of life in capitalist ruins. Princeton University Press
14. Manzini E (2020) Design and politics of nature. DESIS Philosophy Talk #7.1 "Designing as politics (of nature)", 02/19/2020, Politecnico di Milano, Milano
15. Moggridge B, Atkinson B (2007) Designing interactions. MIT Press, Cambridge
16. Norman D (2013) The design of everyday things: revised and expanded edition. Basic Books
17. Lewis SL, Maslin MA (2015) Defining the Anthropocene. Nature 519:171–180. https://doi.org/10.1038/nature14258
18. Verma AK (2018) Ecological balance: an indispensable need for human survival. J Exp Zool India 21:407–409
19. Schramski JR, Gattie DK, Brown JH (2015) Human domination of the biosphere: rapid discharge of the earth-space battery foretells the future of humankind. Proc Natl Acad Sci U S A 112:9511–9517
20. Vitousek PM, Mooney HA, Lubchenco J, Melillo JM (1997) Human domination of earth's ecosystems. Science 277:494–499. https://doi.org/10.1126/science.277.5325.494
21. Johnson CN, Balmford A, Brook BW et al (2017) Biodiversity losses and conservation responses in the Anthropocene. Science 356:270–275
22. Crutzen PJ, Stoermer EF (2000) The "Anthropocene". Glob Chang Newslett 41:17–18
23. Braidotti R, Hlavajova M (2018) Posthuman glossary. Bloomsbury Publishing
24. Ellis EC, Ellis EC (2018) Anthropocene: a very short introduction. Oxford University Press
25. Cook J, Oreskes N, Doran PT et al (2016) Consensus on consensus: a synthesis of consensus estimates on human-caused global warming. Environ Res Lett 11:048002
26. Ellis EC, Ramankutty N (2008) Putting people in the map: anthropogenic biomes of the world. Front Ecol Environ 6:439–447
27. Ellis EC (2015) Ecology in an anthropogenic biosphere. Ecol Monogr 85:287–331
28. McManus K (2018) How anthropo-scenic! Concerns and debates about the age of the anthropocene. In: Anthropocene: Burtynsky, Baichwal, De Pencier, pp 45–56
29. Kothari A, Salleh A, Escobar A et al (2019) Pluriverse. A post-development dictionary. Tulika Books, New Dehli
30. Morin E, Abouessalam S (2020) Cambiamo strada: le 15 lezioni del Coronavirus. Raffaello Cortina

31. Diaz S, Settele J, Brondízio E, Ngo HT, Guèze M, Agard J, Arneth A, Balvanera P, Brauman K, Butchart S, Chan K, Garibaldi LA, Ichii K, Liu J, Subramanian SM, Midgley GF, Miloslavich P, Molnár Z, Pfaff A, Polasky S, Purvis A, Razzaque J, Reyers B, Shin JJ, Visseren-Hamakers IJ, Willis K, Zayas C (2019) The global assessment report on biodiversity and ecosystem services: Summary for policymakers. Intergovernmental Science-Policy Platform on Biodiversity and Ecosystem Services (IPBES), Bonn
32. Forlano L (2016) Decentering the human in the design of collaborative cities. Des Issues 32:42–54
33. Fry T (2017) Remaking cities: an introduction to urban metrofitting. Bloomsbury Publishing
34. Laybourn-Langton L, Rankin L, Baxter D (2019) This is a crisis: Facing up to the age of environmental breakdown. IPPR, London
35. Smith N, Bardzell S, Bardzell J (2017) Designing for cohabitation: naturecultures, hybrids, and decentering the human in design. In: Proceedings of the 2017 CHI conference on human factors in computing systems, pp 1714–1725
36. Mignolo WD (2018) On pluriversality and multipolar world order: decoloniality after decolonization; dewesternization after the cold war. In: Constructing the pluriverse. Duke University Press, pp 90–116
37. Quijano A (2000) Coloniality of power and Eurocentrism in Latin America. Int Sociol 15:215–232
38. Tlostanova M (2017) On decolonizing design. Des Philos Pap 15:51–61
39. Descola P, Pálsson G (1996) Nature and society: anthropological perspectives. Taylor & Francis
40. Latour B (2018) Down to Earth: politics in the new climatic regime. Wiley
41. Haraway D (1988) Situated knowledges: the science question in feminism and the privilege of partial perspective. Fem Stud 14:575–599
42. Pierotti R (2011) Indigenous knowledge, ecology, and evolutionary biology. Routledge, New York
43. Maturana HR, Verden-Zöller G (2008) The origin of humanness in the biology of love. Imprint Academic
44. Lovelock JE, Margulis L (1974) Atmospheric homeostasis by and for the biosphere: the Gaia hypothesis. Tellus 26:2–10
45. Willis A-M (2006) Ontological designing. Des Philos Pap 4:69–92
46. Forlano L (2017) Posthumanism and design. She Ji J Des Econ Innov 3:16–29
47. Giaccardi E, Redström J, Nicenboim I (2025) The making (s) of more-than-human design: introduction to the special issue on more-than-human design and HCI. Hum-Comput Interact 40:1–16
48. Tironi M, Albornoz C, Chilet M (2022) Problematizing human-centred design: notes on planet-oriented design. Diid Disegno Industriale Ind Des (77):12
49. Wakkary R (2021) Things we could design: for more than human-centered worlds. MIT Press
50. Tironi M, Garretón M (2024) Hybrid ecologies of artificial intelligence: prototyping terrestrial practices through a design installation. In: Gray C, Ciliotta Chehade E, Hekkert P, Forlano L, Ciuccarelli P, Lloyd P (eds) DRS2024: Boston, 23–28 June, Boston, USA. https://doi.org/10.21606/drs.2024.1104
51. Borthwick M, Tomitsch M, Gaughwin M (2022) From human-centred to life-centred design: considering environmental and ethical concerns in the design of interactive products. J Responsible Technol 10:100032
52. Lutz D (2022) The life-centred design guide. Damien Lutz
53. Giaccardi E, Redström J (2020) Technology and more-than-human design. Des Issues 36:33–44. https://doi.org/10.1162/desi_a_00612
54. Liu J, Byrne D, Devendorf L (2018) Design for collaborative survival: an inquiry into human-fungi relationships. In: Mandryk R, Hancock M, Perry M, Cox Anna (eds) Proceedings of the 2018 CHI Conference on Human Factors in Computing Systems, pp 1–13. Association for Computing Machinery, New York

55. Brennan A, Lo N (2002) Environmental ethics. In: Zalta EN, Nodelman U (eds) The Stanford Encyclopedia of Philosophy. The Metaphysics Research Lab, Stanford
56. Coccia E (2018) La vita delle piante: metafisica della mescolanza. Società editrice il Mulino
57. Fry T (2009) Design futuring. University of New South Wales Press, Sydney, pp 71–77
58. Mathews F (2011) Towards a deeper philosophy of biomimicry. Organ Environ 24:364–387
59. Coccia E (2021) Métamorphoses. Wiley
60. Latour B (2007) Reassembling the social: an introduction to actor-network-theory. OUP Oxford
61. Law J (2004) After method: mess in social science research. Routledge
62. de la Bellacasa MP (2017) Matters of care: speculative ethics in more than human worlds. University of Minnesota Press
63. Bennett J (2010) Vibrant matter: a political ecology of things. Duke University Press
64. DeLanda M (2016) Assemblage theory. Edinburgh University Press
65. Deleuze G, Guattari F (1988) A thousand plateaus: capitalism and schizophrenia. Bloomsbury Publishing
66. Tsing AL, Bubandt N, Gan E, Swanson HA (2017) Arts of living on a damaged planet: ghosts and monsters of the anthropocene. University of Minnesota Press
67. DiSalvo C, Lukens J (2011) Nonanthropocentrism and the nonhuman in design: possibilities for designing new forms of engagement with and through technology. In: From social butterfly to engaged citizen: urban informatics, social media, ubiquitous computing, and mobile technology to support citizen engagement, p 421
68. Sayes E (2014) Actor–Network Theory and methodology: just what does it mean to say that nonhumans have agency? Soc Stud Sci 44:134–149
69. Bratteteig T, Verne G (2018) Does AI make PD obsolete? Exploring challenges from artificial intelligence to participatory design. In: Proceedings of the 15th participatory design conference: short papers, situated actions, workshops and tutorial, vol 2, pp 1–5
70. Bridle J (2022) Ways of being: animals, plants, machines: the search for a planetary intelligence. Penguin UK
71. Kuijer L, Giaccardi E (2018) Co-performance: conceptualizing the role of artificial agency in the design of everyday life. In: Proceedings of the 2018 CHI conference on human factors in computing systems, pp 1–13
72. Lupetti ML, Bendor R, Giaccardi E (2019) Robot citizenship: a design perspective. Des Semant Form Mov 87:81–89
73. Neuhauser L, Kreps GL (2011) Participatory design and artificial intelligence: strategies to improve health communication for diverse audiences. In: 2011 AAAI Spring symposium series
74. Gatto G, McCardle JR (2019) Multispecies design and ethnographic practice: following other-than-humans as a mode of exploring environmental issues. Sustainability 11:5032
75. Metcalfe D (2015) Multispecies design (Doctoral dissertation). University of the Arts London and Falmouth University, London
76. Tomico O, Wakkary R, Andersen K (2023) Living-with and designing-with plants. Interactions 30:30–34
77. Westerlaken M (2020) Imagining multispecies worlds. Malmö University
78. Lawrence JL, Wiebe SM (2018) Biopolitical disaster. Taylor & Francis
79. Von Essen E, Allen MP (2017) Solidarity between human and nonhuman animals: representing animal voices in policy deliberations. Environ Commun 11:641–653
80. Escobar A (2019) Habitability and design: radical interdependence and the re-earthing of cities. Geoforum 101:132–140
81. Viveiros de Castro E (2012) Cosmological perspectivism in Amazonia and elsewhere. Masterclass Series 1, HAU Network of Ethnographic Theory, Manchester
82. Hall M (2011) Plants as persons: a philosophical botany. Suny Press
83. Haraway D (2016) Staying with the trouble: making kin in the Chthulucene. Duke University Press

84. Morton T (2016) Dark ecology: for a logic of future coexistence. Columbia University Press
85. Tronto JC (1993) Moral boundaries: a political argument for an ethic of care. Psychology Press
86. Escobar A (2019) Thinking-feeling with the earth: territorial struggles and the ontological dimension of the epistemologies of the south. In: Knowledges born in the struggle. Routledge, pp 41–57
87. Popke J (2009) Geography and ethics: non-representational encounters, collective responsibility and economic difference. Prog Hum Geogr 33:81–90
88. Ingold T (2011) Being alive: essays on movement, knowledge and description. Routledge
89. Jain A (2020) Calling for a more-than-human politics. In: Superflux. https://superflux.in/index.php/calling-for-a-more-than-human-politics/. Accessed 18 Dec 2022
90. Chao S, Celermajer D (2023) Introduction: multispecies justice. Cult Polit 19:1–17
91. Aiyadurai A, Chattopadhyay A, Choksi N (2023) Ecological entanglements: affect, embodiment and ethics of care. Orient BlackSwan
92. Ávila M, Steagall MM (2022) Designing for Interdependence: A Poetics of Relating. In: Mortensen Steagall M, Nesteriuk S (eds.) Proceedings of LINK 2022 4th Edition of the International Conference of Practice and Research in Design & Global South Vol.3 No.1, pp 21–24. Bloomsbury Visual Arts, London
93. Akama Y, Light A, Kamihira T (2020) Expanding participation to design with more-than-human concerns. In: Del Gaudio C, Parra-Agudelo L, Clarke R, Saad-Sulonen J, Botero A, Londoño FC, Escadóno P (eds) PDC '20: Proceedings of the 16th Participatory Design Conference 2020 - Participation(s) Otherwise - Volume 1, pp 1–11. Association for Computing Machinery, New York
94. Palmås K, von Busch O (2015) Quasi-quisling: co-design and the assembly of collaborateurs. CoDesign 11:236–249
95. Rice L (2018) Nonhumans in participatory design. CoDesign 14:238–257
96. Tolbert D, Buzzanell PM, Zoltowski CB et al (2016) Giving and responding to feedback through visualisations in design critiques. CoDesign 12:26–38
97. Groutars EG, Kim R, Karana E (2024) Designing living artefacts for multispecies interactions: an ecological approach. Int J Des 18:59–78
98. Head L, Atchison J, Phillips C, Buckingham K (2017) Vegetal politics: belonging, practices and places. Routledge
99. Morton DB, Burghardt GM, Smith JA (1990) Critical anthropomorphism, animal suffering, and the ecological context. Hast Cent Rep 20:S13–S13
100. Abram D (1996) The spell of the sensuous: perception and language in a more-than-human world, 2017th edn. Vintage, New York
101. Nijs G, Laki G, Houlstan R, et al (2020) Fostering more-than-human imaginaries: introducing DIY speculative fabulation in civic HCI. In: Lamas D, Sarapuu H, Šmorgun I, Berget G (eds) NordiCHI '20: Proceedings of the 11th Nordic Conference on Human-Computer Interaction: Shaping Experiences, Shaping Society, pp 1–12. Association for Computing Machinery, New York
102. Rosner DK (2018) Critical fabulations: reworking the methods and margins of design. MIT Press
103. Huffman JL (1992) Do species and nature have rights? Pub Land Law Rev 13:51

Chapter 2
Shifting Perspectives on Plants

Abstract In cities, plants are often considered as passive entities that are distinctly different from humans. This anthropocentric bias has consistently marginalised plant life from ethical, political, and even design considerations. Recent research in plant biology, plant neurobiology, and environmental humanities contests these assumptions, revealing plants as dynamic entities with intricate techniques of adaptation, perception, and communication. These studies highlight forms of distributed intelligence and basic cognition, which rethink plant behaviour as potentially intentional rather than purely mechanical. This chapter integrates scientific insights regarding plant agency to reposition plants as agents within ecological and design systems. It seeks to transcend disciplinary boundaries and emphasise interspecies similarities by formalising and translating specialised biological knowledge into comprehensible concepts. When considered from a systemic perspective, these frameworks could hold the potential to mediate between human and nonhuman concerns, thereby enabling imagination, speculation, and the envisioning of alternative forms of coexistence.

2.1 Introduction

Within the broad spectrum of nonhumans, plants are often regarded as less complex agents than animals [1, 2]. Commonly perceived as mindless and voiceless [3], they are generally believed to lack the capacity for interacting with humans. Unlike animals, which, despite being distinctly different from humans, still show evident responses, plants act on a much slower scale of growth and interaction [4, 5]. Their biological structure limits rapid development and fails to produce visible signals of perception or communication in response to humans. Their sedentary condition further reinforces the perception of "uselessness" from a human-centred perspective, contributing to the exclusion of plant life from ethical and political consideration [1].

This marginalisation may also stem from the differing ways in which various life forms engage with time. Marder [6], through his concept of "ontophytology"—a combination of "ontology" and "phytology" (the study of plants)—emphasises the

F. Vergani, *Reframing Urban Ecologies*, SpringerBriefs in Applied Sciences and Technology, https://doi.org/10.1007/978-3-032-18924-0_2

distinct temporal experience of plants, which contrasts with human perception. Plants exist on a different timescale, characterised by extended lifespans and much slower processes of growth and reproduction. However, despite their immobility and slower pace, plants have evolved complex survival strategies by adapting to their environments and leveraging atmospheric features—as well as interactions with animal species—to disperse their seeds and expand their genetic reach [7–9].

Although plants may seem less complex from an anthropocentric perspective, a wide array of scholars now interpret their behaviours as intelligent [5, 8, 10–12]. Recent studies in Critical Plant Studies (a transdisciplinary field between the environmental humanities, plant sciences, art and aesthetics, philosophy, literature and ethics) have shown that plants can differentiate between positive and negative stimuli, a phenomenon some researchers interpret as a primitive form of cognition [8]. Plants seem to show what is known as "anoetic consciousness", a basic awareness that enables them to perceive stimuli, respond adaptively, and retain rudimentary forms of memory [13]. These signs of sentient-like behaviour, as noted by Baluška & Reber [14], challenge the traditional view that plant responses are entirely mechanical or directed solely by genetics or environmental conditions [15]. As Karban [16] explains, "many different plant tissues have developed conspicuous and directed movements in response to light, gravity and other environmental stimuli [allowing] plants to be more efficient at capturing resources such as light compared with individuals that are prevented from moving." While the notions of plant intelligence and behaviour have been explored in the past, recent scientific advancements have brought renewed significance to these concepts, especially in terms of distributed cognition and the recognition of plants as nonhuman agents [17].

This second chapter draws primarily on literature from the natural sciences that explores plant agency. Its purpose is to assemble and present insights that move the consideration of plants beyond passive or merely decorative features, but as active participants in ecological systems. The knowledge presented here—rooted mainly in natural studies concerning the biology of plants—is intended to enhance plant agency by bridging the scientific understanding of plant biology with speculation, thereby foregrounding the behaviours of nonhumans.

2.2 Rethinking Plant Intelligence and Agency

2.2.1 What Is Intelligence? The Cognitive Turn in Plant Studies

The contentious and misleading recognition of plants as individuals [18] is closely linked to the notion of intelligence, a cognitive capacity traditionally reserved for the animal kingdom [15]. Drawing from Legg and Hutter [19], the biologist Anthony Trewavas [20] defines intelligence as:

> "(a) […] a property that an individual has as it interacts with its environment or environments, (b) [*something*] related to the agent's ability to succeed or profit with respect to some goal or objective, and (c) [*something depending*] on how able the agent is to adapt to different objectives or environments."

Although broadly defined, intelligence is often associated with the concept of "adaptively variable behaviour" [20], a trait that plants appear to possess. The concept of plant intelligence was first theorized by Charles Darwin in the nineteenth century [12], specifically in relation to the role of roots in directing plant movement. However, it is only in recent decades that scientific research has provided compelling evidence, opening up previously unimaginable new possibilities. As Gibson and Baylee [17] note, we are "discovering the multiple tendrils of vegetal being that have emerged from new knowledge that plants have greater sensory capacities than previously thought."

The early 2000s marked the emergence of plant neurobiology, a research field dedicated to understanding how higher plants receive signals from their environment, process this information, and determine responses that enhance their survival [11, 21, 22]. This interdisciplinary field examines how plants perceive, interpret, and respond to environmental stimuli in complex ways that, although not identical, resemble the functions of animal nervous systems. Findings from plant neurobiology suggest that plants function as embodied agents possessing a distributed form of intelligence. They interact with their surroundings through their entire physical structure, using a modular, decentralised system [12, 23].

Unlike animals, plants lack a central nervous system or brain [5, 24]. Their physiological functions are not concentrated in specific organs but are distributed throughout the organism, allowing for decentralised cognition [12, 18]. Plants engage with their environment by responding to both *biotic* and *abiotic* factors to determine optimal strategies for survival [10, 25] (see Chap. 3). Despite the absence of neurons and synapses, they are capable of generating and transmitting electrical signals through specialised cells. These signals travel from root tips to the canopy, facilitating internal communication and coordination among various plant organs [26].

Like all living organisms, plants show intelligence through their behaviour, which Karban [16] defines as "rapid morphological or physiological responses to events, relative to the lifetime of an individual". Expanding on the definition, Gorzelak et al. [23], citing McNickle et al. [27], describe behaviour as "an expression of plant plasticity that is like a decision point, where each choice involves trade-offs that will affect fitness". In this sense, plant behaviour is deeply tied to adaptive success, which is the capacity to modify phenotypes in response to complex and changing environmental conditions [5, 23, 28]. These phenotypic changes can include movement, physiological adjustments, the growth of new tissue, or the shedding of old structures [23]. Karban and Orrock [29] later refined the concept of plant behavior by identifying four key components: (1) a cue or stimulus that conveys information, (2) a process of perception and judgment in which the plant interprets the cue, (3) a decision-making phase where the plant weighs possible responses based on relative costs and benefits, and (4) the resulting action. While movement in response to a stimulus is relatively straightforward to observe, the more remarkable aspect lies in the cognitive-like processes that underlie such responses. Trewavas [5] argues that the ability to choose, decide, and act requires plants to evaluate information from past experiences, suggesting a type of memory. Karban [16] supports this idea, noting that plants appear capable of memory through both conditioning (being shaped by their own experiences) and preconditioning (influenced by the experiences of parent plants). Through such processes, plants can adjust their behavioural responses over time and even across generations, indicating an ability to learn from environmental cues and enhance their chances of survival [4].

Despite considerable empirical evidence, the concept of plant intelligence remains a controversial topic among researchers [8]. One critical challenge faced by scientists advocating for plant intelligence or sentience is the absence of a brain and nervous system, as well as uncertainties surrounding the inner circuits responsible for transmitting electrical signals [5, 24–26]. According to Gibson and Baylee [17], this scepticism arises primarily from a human-centred bias stemming from ignorance, which restricts intelligence to organisms possessing a brain or linguistic capability.

Nevertheless, evidence suggests that plants are purposeful and deliberative beings that actively manage their own existence [22, 30]. They demonstrate remarkable behaviours and actions, not only recognising and responding to immediate environmental cues but also anticipating future conditions and developing strategies rooted in cognitive and conscious processes [15, 16]. These findings prompt fundamental questions regarding our very definition of intelligence: is reasoning, problem-solving, or conscious decision-making truly essential to the concept of intelligence? Given the resilience plants have demonstrated throughout the evolutionary history of our planet, can we genuinely deny them a distinct form of intelligence?

2.2.2 Plant Agency

Granting *agency* to plants—defined by Latour [31] as the "power to act"—recognises them as legitimate and equally significant actors within the Earth's ecosystem. As previously outlined, plants continuously shape and transform the shared environments they, and we, inhabit through their behaviours and actions. From a distinctive philosophical perspective, Coccia [32] refers to plants as the "original designers," noting that "without the oxygen produced by photosynthesis, the Earth's atmosphere would not have changed its internal composition [to become] the most immediate environment for all living things" [32]. This view is echoed by Mancuso [33], who asserts that plants "are extraordinary beings in every respect [as you] do not become the primary source of a planet's life by chance." Yet, despite such compelling insights, this perspective remains neglected. Plants continue to be perceived as "ontologically inferior" [17], lacking equal standing with other life forms. Even before scientific recognition of plant intelligence, various disciplines had already begun to open an ethical discourse around the world of plants, advocating a shift away from viewing them as merely passive or "romantic" [34] entities in need of protection, and instead toward understanding them as active and purposeful agents. This discourse has invigorated debates around plant ethics, challenging dominant human-centred frameworks. From a human perspective, it raises the question: is it ethical to engage with the environment and its diverse inhabitants from an anthropocentric standpoint? From the plant side, a more radical question emerges: can plants themselves be regarded as ethical subjects? [1, 6, 12, 35–39].

In line with these questions, Marder [6] calls for a fundamental shift from utilitarian to ethical cultural values, advocating for the recognition of plants as moral beings in their own right. As Sandler [40] observes, plants are typically valued for their instrumental utility as they are seen as resources that provide food, shade, materials, recreation, aesthetic experiences, and medicine. In this sense, they are often treated as "welfare subjects" [41], entities whose existence contributes to human well-being. Yet, despite their acknowledged influence on human life and society, the intrinsic worth of plants is frequently overlooked [40]. To address this ethical gap, Mathews [42] proposes a bioinclusive environmental ethics. This approach begins with human moral reasoning but expands it to encompass the interests and concerns of nonhuman agents. A particularly compelling ethical framework is again presented by Marder [6], who introduces the concept of "vegetal democracy," an extension of democratic principles to include plants and other nonhumans. This idea emphasises the need to recognise the agency, rights, and ecological interconnectedness of plants within the broader web of life. According to Marder, conventional democracy rooted in human supremacy and the exclusion of nonhumans is inadequate to confront the ecological and ethical crises of our time.

As previously discussed, recognising plants as individuals [15, 18] entails acknowledging their role within the broader collective of life. Just as individuals who contribute meaningfully to society are granted rights that protect their well-being and affirm their value, plants, too, deserve recognition through rights, policies, and practices that reflect their *inherent worth* [40] (in essence, their agency).

In recent years, scholars and science communicators have increasingly advocated for the extension of legal rights to plants [1, 12, 17, 33, 39, 43–47], thereby recognising them as political actors. Broadly speaking, holding a *political role* refers to how individuals or groups participate in or influence political processes and decision-making within a society. This includes actions such as civic engagement and the right to vote, all of which contribute to the functioning of a democracy. When it comes to plants, however, the notion of rights remains highly contested. The concept of rights is traditionally reserved for beings with consciousness and the capacity for moral reasoning [48, 49]. Many, including animal rights advocate Tom Regan [50], claim that rights apply only to sentient entities (humans and animals) and cannot be extended to non-sentient life forms such as plants.

However, some scholars argue that plants deserve moral consideration and protection due to their *inherent worth* as living beings and their profound ecological impact [40]. From this perspective, an agent's political role is tied to its legal status, particularly to its capacity to claim rights within the framework of national or global citizenship. While a few pioneering nations have begun recognising nonhumans within legal frameworks, the vast majority of countries continue to exclude plants from legal representation, a gap widely acknowledged in the literature [1, 6, 12, 33, 35, 37–39]. Although many sources emphasise the urgent need to establish legal rights for plants, two notable works stand out for offering concrete, speculative frameworks. In *The Nation of Plants* [11], botanist Stefano Mancuso envisions a fictional *Bill of Rights* for plants, while journalist Alessandra Viola does the same in *Flower Power* [47]. Both propose sets of articles designed to grant plants symbolic legal recognition and protection. While the data informing these proposals are grounded in scientific research—particularly the findings of plant neurobiology and its revelations about plant intelligence—the articles themselves remain speculative, as they are authored by humans and not directly derived from the plants they seek to represent.

However, while the idea of granting moral standing to plants is gaining attention, academic literature on plant rights remains scarce, especially when compared to the well-established body of work on animal rights [51, 52]. There is a need to recognise the rights of nonhumans, not merely in terms of their utility to humans, but as entities deserving protection from harm [17]. Although countries like Ecuador and Bolivia have taken pioneering steps in granting legal rights to nonhumans, such initiatives remain the exception rather than the rule. Alessandra Viola [47] offers a detailed account of the gradual and limited progress in recognising plants as legal subjects across various national legislations. While global organisations, such as the United Nations, have begun to engage with this issue through forums, policy proposals, and agendas, most Western countries continue to resist extending legal status to nonhumans. In contrast, many Indigenous societies such as the Haida, an indigenous community who have traditionally occupied Haida Gwaii, an archipelago just off the coast of British Columbia (Canada), for at least 12,500 years, have long upheld legal and spiritual frameworks that recognise the rights of plants, animals, and other agents within the ecosystem. As Abbott [53] notes, such recognition is not only ecologically significant but also serves as a means of restoring respect for Indigenous communities, cultures, and knowledge systems that have historically been marginalised or cancelled by Western colonial paradigms.

The political recognition of plants, and, more broadly, of all nonhumans, requires a reimagining of politics as an open, inclusive system that fully integrates ecological thought. The longstanding divide between politics and ecology has played a key role in reinforcing the false dichotomy between humans and *nature* [34, 54, 55]. Concepts rooted in "biopolitics" [56] have given rise to new paradigms of political and ethical engagement, such as "posthuman ethico-politics" [57] and "alterbiopolitics" [58] that emphasise "radical interdependence" [59] among all living agents. These frameworks are grounded in the relational and transformative practice of care, as described in Chap. 1 [58, 60, 61]. The key challenge lies in exploring how these processes of care can reveal the deep interdependence between life forms, whether through institutional policy or grassroots action. When manifested through deliberate practices, these initiatives have the potential to reshape politics on a broader scale.

There is a growing call to rethink our relationship with plants, recognising their agency and acknowledging the impact of human actions on their existence. Building on Chap. 1, the ability to *design* should no longer be viewed as an exclusively human trait. Instead, it should be expanded and understood as a *mode of action* [62], shaped by cognition, memory, and responsiveness to stimuli. In this light, plants, possessing their own sensibilities, languages, and ways of being, many of which remain imperceptible to humans [53], may indeed exert a type of *design movement* [63] or *skill*, shaping and influencing their surrounding environment. As with many ethical struggles, their recognition relies mainly on the support of a civil society willing to speak on behalf of these silent agents. Design, in particular, plays a vital role in this process, both as a medium for representation and as a catalyst for change [64, 65].

2.3 Plant Agency Breakdown: Goals, Components and Behaviours

Research shows that, like all living organisms, plants have evolved to develop specific traits and behaviours that enhance their ability to survive and thrive on Earth [7–9]. For instance, plants have adapted distinct forms, sizes, and specialised organs (such as roots and leaves) that optimise their capacity to catch sunlight, absorb water, and extract nutrients from the soil [4, 16, 23, 26]. These adaptive behaviours are not confined to long-term evolutionary processes but also manifest daily, as plants respond dynamically to environmental changes [5, 23, 28]. Such responses reflect more than simple biological reactivity: they represent complex, intentional movements and interactions with other nonhumans. These complex behaviours are manifested through actions [62] that suggest an intentionality in how plants move through space and engage with other nonhuman agents.

Taken together, these behaviours lend support to the increasingly discussed hypothesis that plants are not only intelligent [5, 66] but also sentient beings [14, 24, 25], capable of deliberate behaviour and action that aligns with the broader concept of design articulated in Chap. 1. If we understand design as an intentional process involving "negotiations, exchanges, articulations, and nourishments" [67], then it becomes conceivable that plants, too, engage in a form of *design-driven movement*, actively shaping and influencing their surrounding environment.

Before delving further into this contentious topic, it is essential to clarify the scientific foundations that support the concept of plant agency. The current body of knowledge presented relies on highly specialised and complex data, which can be challenging to interpret and translate across disciplinary boundaries. The work presented in this volume begins with this initial phase of data collection aimed at better understanding the *modus operandi* of plants, thereby enabling a more fluid engagement with this particular body of knowledge. Accordingly, the chapter seeks to systematise emerging and, in many cases, still ambiguous or evolving forms of knowledge through the review of studies drawn from disciplines such as biology, botany, and plant neurobiology. These data are assembled to support the reconstruction of plant behaviour from a systemic perspective, grounded in principles of relation and causality, and encompassing the life cycles of vegetal agents, their goals, and the specific mechanisms through which these are pursued.

The chapter does not simply aim to document plant behaviours for the sake of dissemination; rather, its purpose is to translate these insights into a systemic understanding that allows the tracing of vegetal behaviour as a *process*. This, in turn, makes it possible to highlight the affinities between this way of behaving and that of all terrestrials, including humans. Although this may appear to be a technical exercise, it represents a crucial step in moving beyond a romanticised view of *nature* as a chaotic entity, and towards the generation of multispecies frameworks grounded in the democratisation of agency among all participating species. While this empirically driven approach may at times seem to flatten interspecies differences, its intention is precisely the opposite: to demonstrate that a shared ground of relation between humans and nonhumans is scientifically conceivable when examined through the lens of common structural principles. Thus, despite the potential criticisms such a technical interpretation may invite, the effort to translate the unique and distinctive behaviours of plants into a coherent systemic model is of fundamental importance to the aims of this work.

As explained earlier, the knowledge presented in this section is derived from a process of data gathering conducted through both scientific and popular sources. Its inclusion is essential for grounding the basic concepts of the natural sciences and for enabling the subsequent steps and design-oriented interpretations developed in the following chapters. To establish the smoothest possible progression of ideas, this subsection organises the material around three core elements: *goals*, *components*, and *behaviours* (Fig. 2.1).

2.3.1 Goals

Plants are multicellular organisms classified in the kingdom Plantae, manifesting in many forms across nearly all ecosystems on Earth. Their distinguishing features include cellulose-based cell walls, chloroplasts for photosynthesis, and the ability to synthesise their own food through this process [68]. They have intricate life cycles and are capable of reproducing both sexually and asexually. These life cycles, much like those of all agents [69], play a fundamental role in shaping behaviour. Among both living and non-living entities, few elements are as universally shared as the life cycle itself, which provides a framework for understanding how an agent operates in pursuit of a specific *goal*. This sub-section outlines the main stages of a plant's life cycle. Knowledge of these stages is well established and commonly introduced at primary and secondary school levels. Accordingly, the information presented

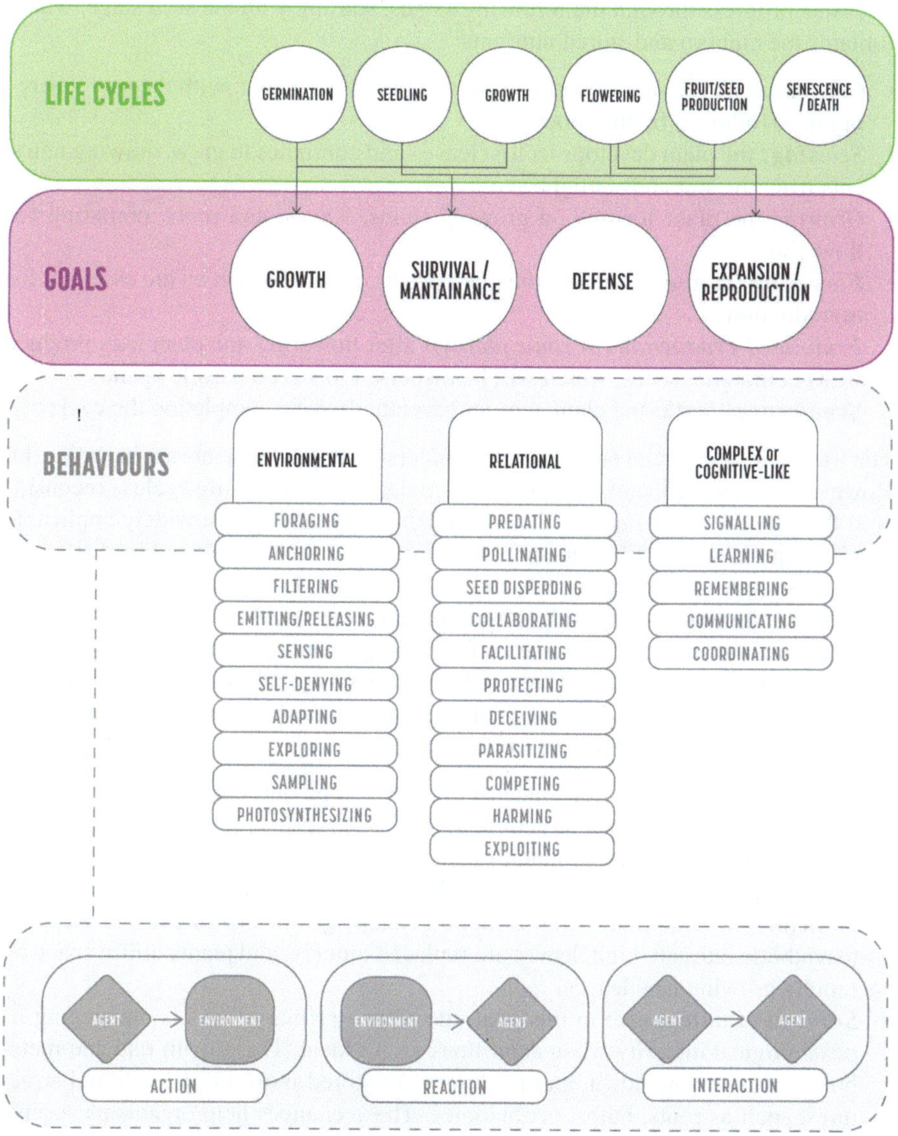

Fig. 2.1 Plant's *life cycles*, *goals*, and *behaviours*

here has been drawn primarily from textbooks and online scientific sources. It is offered in a simplified form, serving as a foundational reference for the subsequent sections of this work. For the sake of clarity, throughout this book, only plant species that originate from seed will be considered, excluding those reproduced through cloning (such as grafting or cuttings) or through spores.

Plants progress through the following stages, starting with the seed stage, which contains the embryo and stored nutrients:

- ***Germination*:** the seed absorbs water and begins to sprout, with the root emerging first, followed by the shoot.
- ***Seedling*:** the plant develops its first leaves and continues to grow, drawing nutrients from the soil and sunlight.
- ***Growth*:** the plant focuses on growing stems, leaves, and roots, preparing for flowering.
- ***Flowering (in angiosperms)*:** the plant produces flowers, which are essential for reproduction.
- ***Fruit/Seed Production (in some plants)*:** after flowering, the plant may produce fruits containing seeds, which will be dispersed to start the cycle again.
- ***Senescence/Death*:** the plant ages and eventually dies, completing the cycle.

The life cycle is a crucial entry point for understanding how plants act in the world. However, in framing the idea of plant-driven *design skills*, the life cycle is reconsidered through the lens of *goals*. To make this interpretation more widely applicable and intelligible across different species, some stages have been combined (such as those relating to growth or reproduction). In contrast, others (most notably death) have been intentionally left aside. These are not rigid stages that occur in sequence but overlapping priorities that rise or fall in importance depending on the plant's age, the conditions it faces, and the evolutionary strategy it has inherited. Therefore, the final *goals* have been clustered into four categories: *growth*, *survival/maintenance*, *defence*, and *expansion/reproduction*.

- ***Growth*:** growth is the first stage of plant development, during which the plant builds up its physical and structural ability to survive. From the moment a seed germinates, a significant amount of metabolic energy is invested in producing photosynthetically active leaves, developing root systems to access water and nutrients, and extending stems to support canopy growth. During this phase, photosynthetic rates are high, leaves are replaced quickly, and plants utilise space by rapidly growing new leaves.
- ***Survival/Maintenance***: in this goal, the plant's primary focus is on keeping its physiological integrity when conditions aren't ideal. The growth rate and metabolic process slow down, and resources are stored more in underground structures, such as roots, bulbs, or rhizomes. These changes help organisms survive when there are abiotic stresses, such as drought, low temperatures, or a lack of nutrients.
- ***Defence***: defence is performed to protect against biotic threats, such as herbivory, pathogenic infections, and mechanical damage. Structural defences can involve the formation of lignified tissues, thick bark, spines, or sclerophyllous leaves. Chemical defences can affect the production of secondary metabolites that repel pests. Plants often form mutually beneficial relationships with animals, giving them flowers or shelter in exchange for protection.

- ***Expansion/Reproduction***: expansion encompasses the processes of reproduction and dispersal through which a plant ensures the continuation of its genetic lineage. This could involve sexual reproduction through the production of flowers, fruits, and seeds, often with specialised adaptations for dispersal that utilise abiotic vectors, such as water and wind, or biotic agents, like animals.

2.3.2 Components

Recognising plants as agents means understanding that their biological structures are not passive: they work actively to help the plant fulfil its various goals. Much like humans and animals, plants rely on a set of components (organs) that allow them to grow, survive, defend themselves, and keep functioning over time. For this inquiry, these components are grouped into three main clusters: roots, stems, and leaves. In order to maintain a scope broad enough to be applicable across a wide range of plant species, other components, such as branches, flowers, seeds, and fruits, are referred to only briefly. Not every plant has all these organs, yet each plays an important role when considered in relation to the behaviours we observe. As with the knowledge presented in the previous subsection, the primary source for the data provided here is the Encyclopædia Britannica.

- ***Roots***: roots are the underground structures that anchor a plant in place, drawing in the water and nutrients it needs to survive. Usually hidden beneath the soil, they are made up of specialised tissues designed to absorb moisture and minerals as efficiently as possible. Unlike stems, roots do not have leaf scars or buds; instead, they are protected at their tip by a small cap and branch internally rather than from outer buds. In many species, this root cap is remarkably dynamic. Beyond shielding the delicate growing tip, it works almost like a sensory organ. It can detect and respond to a wide range of environmental signals, including gravity, touch, low phosphate levels, pockets of nutrient-rich soil, humidity, salt stress, and even light [5]. Through this sophisticated sensing ability, roots take an active, responsive role in guiding the plant's behaviour and shaping how it adapts to its surroundings.
- ***Stem:*** the stem is the plant's primary support system. It keeps the plant upright and is also the primary means by which water, nutrients, and sugars circulate throughout the plant. The stem is composed of various types of tissue and connects the roots and leaves, letting resources move up and down as needed. In trees, the stem is referred to as the trunk, which is composed of several distinct layers: the outer bark, which provides protection; the cambium, a thin layer of actively dividing cells responsible for growth; and the xylem, which conducts water and minerals from the roots to the rest of the plant. Together, these structures ensure both the structural integrity and physiological functioning of the plant.

- ***Leaves***: leaves are the primary organs responsible for photosynthesis in plants. Usually flat and thin, they are shaped to capture as much sunlight as possible while enabling efficient gas exchange. Their internal structure includes the mesophyll, a tissue packed with chloroplasts, the tiny cellular *factories* where photosynthesis takes place. By converting sunlight, carbon dioxide, and water into energy-rich sugars, leaves fuel the plant's growth and sustain its life functions.

Some plants include other components, such as:

- ***Branches (tree)***: branches are the secondary stems of a tree that grow outward from the trunk and support the leaves.
- ***Flowers***: some plants (angiosperms) produce flowers, which are specialised reproductive structures that produce seeds. Flowers are typically colourful and attract pollinators, which help to fertilise the flowers and produce seeds.
- ***Fruits***: fruits are the structures that contain and disperse seeds. Fruits can be edible, like apples or bananas, or inedible, like pinecones or acorns.

2.3.3 Behaviours: Actions, Reactions, Interactions

Plants interact with the surroundings through a range of complex behaviours [8, 16, 23, 26, 29, 66, 70–76]. These behaviours are typically expressed as changes in morphology or physiology, a phenomenon known as *phenotypic plasticity*, in response to environmental stimuli such as chemical signals, light, and water. Plant behaviour can be understood as a form of decision-making, where each response involves trade-offs that ultimately affect the plant's fitness and survival [16, 23, 27]. The manifestation of this plasticity takes the form of specific actions carried out by the plant's components, as previously listed, in relation to their respective *goals*. However, describing action solely as a manifestation of behaviour is insufficient, given that plants, like all agents, exist within a constant web of relations with other entities. For this reason, actions have been distinguished from two additional types of behaviour that inherently involve external agents: reactions and interactions.

- ***Action***: intentional or autonomous processes often directed toward survival, growth, reproduction, or adaptation. They may occur without external triggers.
- ***Reaction***: immediate and specific responses to external stimuli. Often involuntary or passive, depending on the agent.
- ***Interaction***: ongoing or reciprocal relationships between agents or between an agent and its surroundings, often involving both actions and reactions from both sides.

Recognising that actions, reactions, and interactions unfold within a complex web of relationships, it is useful to introduce an additional level of distinction to describe these behaviours more effectively. In fact, some plants' behaviours might occur individually, shaped mainly by the surrounding environment, while encounters with other agents might prompt others. From a human perspective, these relational behaviours might even appear positive or negative, depending on the dynamics involved. As mentioned earlier, certain behaviours also seem to be guided by forms of cognition, an area of research that is still developing but increasingly compelling. For these reasons, the most recognisable plant actions, reactions, and interactions can be grouped into two main categories: *environmental behaviours*, which happen within the plant's immediate context, and *relational behaviours*, which emerge through contact with other agents. A further note is made for those behaviours that are more complex or show signs of cognitive processing.

- ***Environmental***: these behaviours refer to direct, often physiological or developmental responses by a plant to abiotic factors, such as light, temperature, and soil. These responses are typically automatic and limited to a specific area, showing how organisms have changed over time through natural selection rather than through conscious control. Some of the environmental actions, reactions, interactions include foraging, anchoring, filtering, emitting/releasing, sensing, self-denying, adapting, exploring, sampling, and photosynthesising.
- ***Relational***: these behaviours include how plants interact with other agents and ecosystems. Such interactions are constantly evolving and depend on the specific situation. They often involve two-way influences and feedback loops. This category includes predation, pollinating, seed dispersal, collaboration, facilitation, protection, deception, parasitism, competition, and harm.
- ***Complex or cognitive-like***: the notion of cognitive-like behaviour in plants refers to sophisticated processing of information that resembles decision-making, memory, learning, or problem-solving. As previously articulated, though plants lack brains, research has demonstrated remarkable adaptive behaviours that suggest information integration and prioritisation. Some of the complex or cognitive-like skills include signalling, learning, remembering, communicating, and coordinating.

2.4 Conclusions

This plant breakdown was driven by the challenge of engaging with complex and specialised knowledge. It aimed to formalise and, at the same time, simplify scientific insights from the field of natural sciences, particularly those informed by recent discoveries regarding plant intelligence. Special focus was placed on identifying the deliberate behaviours of plants that might indicate intention or purpose, potentially rooted in a form of cognition.

What is particularly interesting to note is that these plant behaviours are closely tied to specific components and goals, like in the animal kingdom. From a more-than-human perspective, it is therefore interesting to trace these *interspecies similarities*, bringing into the same framework elements that are usually regarded as fundamentally different. The motivation behind this inquiry stems from a gap in the way nonhumans are addressed in much of the design literature. Often, references to plant biology remain surface-level, lacking the depth and nuance found in scientific research. This gap arises from the divergent aims, techniques, and methodologies of the two disciplines; thus, a more precise and comprehensive understanding of plant behavioural patterns is necessary. While plant experts typically endorse such data, design researchers need to be familiar with it, particularly to facilitate its conversion into frameworks and systemic considerations that are fundamental to design practice. Understanding how other agents behave may help us shape current design approaches, opening the way to imagining, speculating, and fabulating new ways of coexisting.

References

1. Pellegrino G, Di Paola M (2019) Etica e politica delle piante. DeriveApprodi
2. Silvertown J, Gordon DM (1989) A framework for plant behavior. Annu Rev Ecol Syst 20:349–366
3. Plumwood V (2002) Environmental culture: the ecological crisis of reason. Psychology Press
4. Segundo-Ortin M, Calvo P (2019) Are plants cognitive? A reply to Adams. Stud Hist Philos Sci A 73:64–71
5. Trewavas A (2017) The foundations of plant intelligence. Interface Focus 7:20160098
6. Marder M (2013) Plant-thinking: a philosophy of vegetal life. Columbia University Press
7. Huot OB, Nachappa P, Tamborindeguy C (2013) The evolutionary strategies of plant defenses have a dynamic impact on the adaptations and interactions of vectors and pathogens. Insect Sci 20:297–306
8. Kumar A, Memo M, Mastinu A (2020) Plant behaviour: an evolutionary response to the environment? Plant Biol 22:961–970. https://doi.org/10.1111/plb.13149
9. Vacchiano G (2019) La resilienza del bosco: storie di foreste che cambiano il pianeta. Mondadori, Milano
10. Gagliano M, Abramson CI, Depczynski M (2018) Plants learn and remember: lets get used to it. Oecologia 186:29–31
11. Mancuso S (2019) La nazione delle piante. Gius. Laterza & Figli Spa
12. Viola A, Mancuso S (2015) Verde brillante. Sensibilità e intelligenza del mondo vegetale. Giunti Editore, Firenze
13. Chamovitz D (2020) What a plant knows: a field guide to the senses: updated and expanded edition. Scientific American/Farrar, Straus and Giroux
14. Baluška F, Reber A (2019) Sentience and consciousness in single cells: how the first minds emerged in unicellular species. BioEssays 41:1800229
15. Segundo-Ortin M, Calvo P (2022) Consciousness and cognition in plants. WIREs Cogn Sci 13:e1578. https://doi.org/10.1002/wcs.1578
16. Karban R (2008) Plant behaviour and communication. Ecol Lett 11:727–739
17. Gibson P, Baylee B (2018) Covert plants: vegetal consciousness and agency in an anthropocentric world. Punctum Books
18. Calvo P (2007) The quest for cognition in plant neurobiology. Plant Signal Behav 2:208–211. https://doi.org/10.4161/psb.2.4.4470
19. Legg S, Hutter M (2007) A collection of definitions of intelligence. Front Artif Intell Appl 157:17
20. Trewavas T (2016) Plant intelligence: an overview. BioScience 66:542–551
21. Brenner ED, Stahlberg R, Mancuso S et al (2006) Plant neurobiology: an integrated view of plant signaling. Trends Plant Sci 11:413–419

22. Mancuso S (2017) Plant revolution. Le piante hanno già inventato il nostro futuro Firenze, Milano: Guiuti Editore
23. Gorzelak MA, Asay AK, Pickles BJ, Simard SW (2015) Inter-plant communication through mycorrhizal networks mediates complex adaptive behaviour in plant communities. AoB Plants 7:plv050
24. Calvo P, Sahi VP, Trewavas A (2017) Are plants sentient? Plant Cell Environ 40:2858–2869
25. Pelizzon A, Gagliano M (2015) The sentience of plants: animal rights and rights of nature intersecting. Austl Anim Protect Law J 11:5
26. Hedrich R, Salvador-Recatalà V, Dreyer I (2016) Electrical wiring and long-distance plant communication. Trends Plant Sci 21:376–387
27. McNickle GG, Clair CCS, Cahill JF Jr (2009) Focusing the metaphor: plant root foraging behaviour. Trends Ecol Evol 24:419–426
28. Reid CR, Garnier S, Beekman M, Latty T (2015) Information integration and multiattribute decision making in non-neuronal organisms. Anim Behav 100:44–50
29. Karban R, Orrock J (2018) A judgment and decision-making model for plant behavior. Ecology 99:1909–1919. https://doi.org/10.1002/ecy.2418
30. Trewavas A (2014) Plant behaviour and intelligence. OUP Oxford
31. Latour B (2017) Facing Gaia: eight lectures on the new climatic regime. Wiley
32. Coccia E (2021) Métamorphoses. Wiley
33. Mancuso S (2019) Plant intelligence. In: Trees. Fondation Cartier pour l'art contemporain, Paris, pp 48–57
34. Morton T (2007) Ecology without nature: rethinking environmental aesthetics. Harvard University Press
35. Hall M (2011) Plants as persons: a philosophical botany. Suny Press
36. Huffman JL (1992) Do species and nature have rights? Pub Land Law Rev 13:51
37. Morton T (2016) Dark ecology: for a logic of future coexistence. Columbia University Press
38. Morton T (2017) Humankind: solidarity with nonhuman people. Verso Books
39. Tanasescu M (2014) Rethinking representation: the challenge of nonhumans. Aust J Polit Sci 49:40–53
40. Sandler R (2018) Is considering the interests of plants absurd? In: Plant ethics. Routledge, pp 40–50
41. Višak T (2018) Utilitarian plant ethics. In: Plant ethics. Routledge, pp 30–39
42. Mathews F (2011) Towards a deeper philosophy of biomimicry. Organ Environ 24:364–387
43. Ezra O (2017) The rights of nonhumans: from animals to silent nature. Law Ethics Hum Rights 11:285–304
44. Fry T (2020) Defuturing. A new design philosophy. Bloomsbury, London
45. McGrath SJ (2019) Why political ecology cannot let go of nature. Anal Herm 10
46. Rogers N, Maloney M (2017) Law as if earth really mattered. Taylor & Francis
47. Viola A (2020) Flower power. Le piante e i loro diritti. Passaggi Einaudi
48. Callicott JB (1989) In defense of the land ethic: essays in environmental philosophy. Suny Press
49. Taylor PW (1986) Respect for nature: a theory of environmental ethics, 2011th edn. Princeton University Press
50. Regan T, Singer P (1989) Animal rights and human obligations. Prentice-Hall
51. Cochrane A (2018) Sentientist politics: a theory of global inter-species justice. Oxford University Press
52. Donaldson S, Kymlicka W (2011) Zoopolis: a political theory of animal rights. Oxford University Press
53. Abbott S (2021) Approaching nonhuman ontologies: trees, communication, and qualitative inquiry. Qual Inquiry 27:1059–1071
54. Latour B (2004) Politics of nature. Harvard University Press
55. Latour B (2018) Down to Earth: politics in the new climatic regime. Wiley
56. Lazzarato M (2002) From biopower to biopolitics. Pli Warwick J Philos 13:1–6
57. Braidotti R (2013) Posthuman. Polity Press

58. de la Bellacasa MP (2017) Matters of care: speculative ethics in more than human worlds. University of Minnesota Press
59. Escobar A (2018) Designs for the pluriverse: radical interdependence, autonomy, and the making of worlds. Duke University Press
60. Lawson V (2007) Geographies of care and responsibility. Ann Assoc Am Geogr 97:1–11
61. Tronto JC (1993) Moral boundaries: a political argument for an ethic of care. Psychology Press
62. Rice L (2018) Nonhumans in participatory design. CoDesign 14:238–257
63. Willis A-M (2006) Ontological designing. Des Philos Pap 4:69–92
64. Forlano L (2016) Decentering the human in the design of collaborative cities. Des Issues 32:42–54
65. Manzini E (2020) Design and politics of nature. DESIS Philosophy Talk #7.1 "Designing as politics (of nature)", 02/19/2020, Politecnico di Milano, Milano
66. Falik O, Mordoch Y, Quansah L et al (2011) Rumor has it…: relay communication of stress cues in plants. PLoS One 6:e23625
67. Westerlaken M (2020) Imagining multispecies worlds. Malmö University
68. Singh G (2019) Plant systematics: an integrated approach. CRC Press
69. Werner E (1988) Size, scaling, and the evolution of complex life cycles. In: Size-structured populations: ecology and evolution. Springer, pp 60–81
70. Callaway RM (2007) Positive interactions and interdependence in plant communities. Springer
71. Delory BM, Delaplace P, Fauconnier M-L, Du Jardin P (2016) Root-emitted volatile organic compounds: can they mediate belowground plant-plant interactions? Plant Soil 402:1–26
72. Gniazdowska A, Bogatek R (2005) Allelopathic interactions between plants. Multi site action of allelochemicals. Acta Physiol Plant 27:395–407
73. Kegge W, Pierik R (2010) Biogenic volatile organic compounds and plant competition. Trends Plant Sci 15:126–132
74. Rojo E, Solano R, Sánchez-Serrano JJ (2003) Interactions between signaling compounds involved in plant defense. J Plant Growth Regul 22:82–98
75. Tabassum S, Ossola A, Manea A et al (2020) Using ecological knowledge for landscaping with plants in cities. Ecol Eng 158:106049. https://doi.org/10.1016/j.ecoleng.2020.106049
76. Williams NSG, Hahs AK, Vesk PA (2015) Urbanisation, plant traits and the composition of urban floras. Perspect Plant Ecol Evol Syst 17:78–86. https://doi.org/10.1016/j.ppees.2014.10.002

Chapter 3
Acknowledging Interspecies Relationships with the *Relational Atlas*

Abstract Plant behaviours, goals, and actions are grounded in biological processes, yet they remain inextricably connected to the environments and other species with which plants continuously interact. Drawing on Willis's concept of design's "double movement" (2006), the chapter underscores the importance of recognising plants—and, by extension, other nonhumans—as active participants that are shaped by design interventions and capable of responding to them. Building on this premise, the chapter introduces a set of plant *strategies* that align with the previously articulated goals. These *strategies* form the foundation of the *Relational Atlas*, a framework that positions relationships as the primary principle of multispecies interactions. Although conceptual, the *Atlas* encourages designers to move beyond anthropocentric assumptions, to consider the needs and temporalities of nonhuman life, and to envision more equitable and reciprocal modes of coexistence.

3.1 Introduction

The previous chapter focused on gathering and systematising strictly scientific data concerning plant agency by focusing on their *goals*, *components*, and *behaviours* in relation to their biology. Since these considerations concern not only the plant itself but also its surrounding environment—as well as the agents that inhabit it—it is evident that mutualistic interactions influence the relationships. This volume explicitly references the concept of the "double movement" of design, as presented by Willis [1] and elaborated upon in Chap. 1. This concept identifies a specific relationship between those who design and those who are impacted by design. Understanding this *creative force* encourages us to acknowledge plants and other nonhumans as active participants, considering that their presence constantly influences us in the environments we inhabit. Yet in design practice, we tend to rely almost exclusively on human needs and goals, rarely acknowledging the otherness that is deeply entangled with our lives. From a more-than-human perspective, however, design might help us frame the behaviours and actions of plants as comparable to those of humans.

F. Vergani, *Reframing Urban Ecologies*, SpringerBriefs in Applied Sciences and Technology, https://doi.org/10.1007/978-3-032-18924-0_3

At this point, it is critical to consider their relational dimension from a more *strategic* perspective. This is especially true in the context of participatory design, discussed in further detail in Chap. 5, where the relational aspect becomes even more central. Starting from this assumption, and building on the insights previously collected, this chapter outlines a series of *strategies* that reveal how plants navigate the complexities of life. These *strategies* are proposed as a means of strengthening plant agency and providing human communities with practical guidance for incorporating plant perspectives—albeit mediated through human biases and interpretations—into urban design and planning processes.

3.2 Relationships in Multispecies Contexts

3.2.1 Plants and the Others

Plants, like all other organisms, rely on other agents to perform actions and tasks that they cannot complete on their own [2]. Their behavioural changes, both in scope and intensity, can substantially reshape the ecology of a given situated community, influencing all surrounding agents [3]. In their search for light and nutrients, plants actively move through space, extending roots and leaves to probe the soil and air [4]. This foraging behaviour involves multiple strategies to achieve their goals by connecting with environmental features. As noted in Chap. 2, while discussing intelligence, plants communicate both *above* [5–9] and *below* ground [3, 10–12] using electrical, hydraulic, and chemical signals to coordinate within their own bodies, with other plants, and to establish strong interspecies relationships [13–15]. Their organs (roots, stems, and leaves) emit volatile organic compounds (VOCs) that serve as external signals to communicate with neighbouring agents. While internal communication is primarily linked to foraging, external communication can be understood as a strategy for broader interactions and engagement. As Karban [4] notes, plants release complex volatile blends into the air to achieve specific objectives, thereby providing cues to other plants, as well as to animals and bacteria (whether parasites or predators). Whether the goal is reproduction (e.g., pollination) or defence against external threats, plants use their bodies to send VOCs, transmitting messages to surrounding life forms. Below-ground communication presents unique challenges, as roots remain hidden in the soil, making their study and interpretation difficult for humans [16]. Traditionally, roots have been understood primarily in terms of their structural role, which is to anchor the trunk and canopy, and their function in absorbing water and nutrients [17]. However, their extraordinary capacities related to plant intelligence have only recently gained wider recognition. As early as 1896, Darwin, in *The Power of Movement in Plants*, suggested additional roles for roots, identifying the root cap (a specialised tissue at the tip of the root [18]) as a dynamic sensory organ. He described it as a "delocalized brain" capable of perceiving and processing multiple signals [19]. In this sense, roots can be understood as a control centre that probes the soil for resources, nutrients, and information.

This below-ground interaction, primarily dedicated to plant–plant communication, is further enhanced by the presence of fungi. Through the formation of mycorrhizal networks (MNs), "fungi and plants are joined to one another by an underground hyphal network" [20], exchanging soil-foraged nutrients. Moreover, MNs shape the survival, growth, physiology, health, competitive ability and behaviour of the organisms they connect [3], making them one of the most extensive and powerful mutualistic symbioses on Earth. Through this system, plants exchange resources and support one another, while also competing, acting like *hackers* that can redirect nutrients and resources from neighbouring agents [4].

Functioning as a kind of collective brain, the MN allows plants to access and share information, enhancing their capacity to adapt and respond [21]. This type of *collective* or *swarm intelligence* [19, 22], also observed in the animal kingdom, enables seemingly simple individuals to form highly dynamic, coordinated systems that can better resist environmental challenges. In forests, this is particularly evident: interconnected plants demonstrate both mutual aid and competition, collectively strengthening the resilience of the ecosystem [3, 23]. Although the MN is now widely recognised within the scientific community, much about its mechanisms remains unknown [6, 19], with ongoing research continuing to unravel its complexity.

3.2.2 Plant-Human Interaction

Although scientific studies have classified plant interactions in terms of behaviours, it is still unclear and debated how these relationships might be approached from a human standpoint. While research has been instrumental in advancing our understanding—and as Hall [24] observes, "plants and humans share a basic, ontological reality as perceptive, aware, autonomous, self-governed and intelligent beings"—the medium of communication between the two realms remains blurred. As outlined in earlier chapters, this knowledge gap is primarily rooted in the historically marginal role assigned to plants. Across time, they have been perceived less as agents acting with purpose and more as resources to be conquered, domesticated, and exploited for economic gain [25]. This perspective, which is intensified in more industrialized civilizations, is reinforced by the cognitive bias known as "plant-blindness" [26], or "plant awareness disparity" [27], described as the inability of a person to notice plants and/or appreciate their significance beyond an anthropocentric return (as, for example, the improvement of the physical and mental wellbeing of humans).

When considering human–plant interactions (and, by extension, human–animal ones), it is essential to recognise the limitations inherent in interspecies relationships, mainly the absence of a shared mode of communication, or a common language. As Kulick [28] observes, language represents "only a tiny band on a vastly broader spectrum of communication," suggesting that interaction may also occur through images, scents, emotions, sensations, and feelings. In everyday life, however, humans and plants engage with one another constantly and continuously [29], as plants play a fundamental role in shaping the human experience [30]. From the human perspective, DelSesto [25] outlines a wide range of human–plant interactions, spanning peaceful coexistence, tactile immersion, and physical exertion. People engage with plants in diverse environments (public urban spaces, homes, and forests) while also preparing plant-based foods and products, touching, examining, and working with plant materials, whether for professional purposes or personal enjoyment [25].

From a nonhuman perspective, Rice [31] suggests that nonhumans interact with humans through three primary modes: *substitution*, *mediation*, and *communication*. Substitution refers to instances in which a nonhuman replaces human action (for example, a hedge grown to shield a garden from prying eyes). Mediation occurs when a nonhuman facilitates human activity (such as a large rock serving as a seat). Communication, meanwhile, involves a nonhuman conveying information through its external features (for instance, the changing colour of hydrangeas depending on the level of aluminium in the soil) [31]. Although these methods outline potential types of interaction, it remains difficult to envision a plant–human relationship comparable to interactions between humans, or even between humans and animals. From a human point of view, the peculiar biological structure of plants hinders their ability to establish significant interspecific relationships that mirror the complexity of intraspecific communication. As Abrams [32] points out, humans cannot "precisely experience the living sensations of another form."

However, this limitation arises more specifically in Western contexts. As noted in Chap. 1, alternative perspectives, such as "animism" [33], recognise relationships between humans, plants, and other nonhumans [34–36], opening new horizons for novel forms of interspecies connection. Within this framework, Brazilian anthropologist Eduardo Viveiros de Castro [37] introduced the concept of *perspectivism* as a distinctive way of understanding the relationships among life forms. According to de Viveiros de Castro, humans and nonhumans are inherently different, yet all possess the capacity for a point of view (or *perspective*) that shapes their unique understanding of the world. He further contends that the human–*nature* relationship is not rooted in domination, but in coexistence, where diverse beings can interact with and comprehend one another.

3.3 The *Relational Atlas*

The previous chapter investigated plant behaviours through the lens of biology by identifying their specific *goals* and *components*. Yet plants never act in isolation; they are always involved in interspecific relationships, forming what Coccia [38] describes as "social bonds", connections that may be advantageous at times and limiting or even harmful at others. Recognising this relational dimension is essential for moving beyond the perceived divide between humans and nonhumans. To support humans in acknowledging the many patterns and dynamics they share with plant life, it becomes necessary to establish a framework that can map these relationships, making them both visible and intelligible within design discourse and practice.

As a matter of fact, the world "cannot simply be viewed and addressed anthropocentrically, or just in terms of 'the natural'" [39] but must instead be understood through the web of relationships among its various constituent agents. This perspective also draws on Barad's [40] concept of "diffraction," which examines how phenomena arise through mutually constitutive intra-actions and inter-actions, rather than as separate, pre-existing entities. Diffraction offers a way of reading insights through one another, revealing how patterns of difference are formed and re-formed, and how design decisions, species behaviours, and social dynamics collectively shape the shared urban ecosystem. Consequently, the diagram shown in Chap. 2, which considered the plant as an individual agent with its *goals*, *components,* and *behaviours*, is further elaborated and expanded here, providing the basis for the development of a *Relational Atlas* (RA) that integrates all agents and their behaviours into a unified system (Fig. 3.1).

From a more-than-human perspective, it is crucial to explore the relational dimension of plants through a more strategic lens. As discussed in Chap. 1 through the concept of the "double movement" [1], design is inherently relational: it entails both an intentional action by the designer and a corresponding effect on the being or entity impacted. Indeed, however speculative, this viewpoint can help us identify behaviours and actions comparable to those of the human species, almost as if they were guided by forms of design-like agency. Recognising this "creative force" might help us to acknowledge the active role of nonhumans, especially since their presence constantly impacts us in our personal and public spaces. Yet, in design practice, we continue to refer primarily to human needs and objectives, rarely considering other forms of alterity that are deeply intertwined with our lives.

The RA, presented here in its current developmental form, positions plants alongside all other agents as fundamentally relational organisms. In doing so, it invites a rethinking of design, and particularly the urban design of public space, as a process shaped by encounters among agents and the negotiations that emerge from these interactions.

3.3.1 *The* Atlas's *Features*

The RA is built on the premise that all agents share the same capacity for action (agency). Instead of following a traditional top-down, pyramidal hierarchy, it organises information in a circular structure: agents are positioned in a ring, while the core contains information about the relationships connecting them. As in the previous chapter, all data included in the *Atlas* have been extensively gathered from literature in the natural sciences. Although the various features (*agents*, *strategies*,

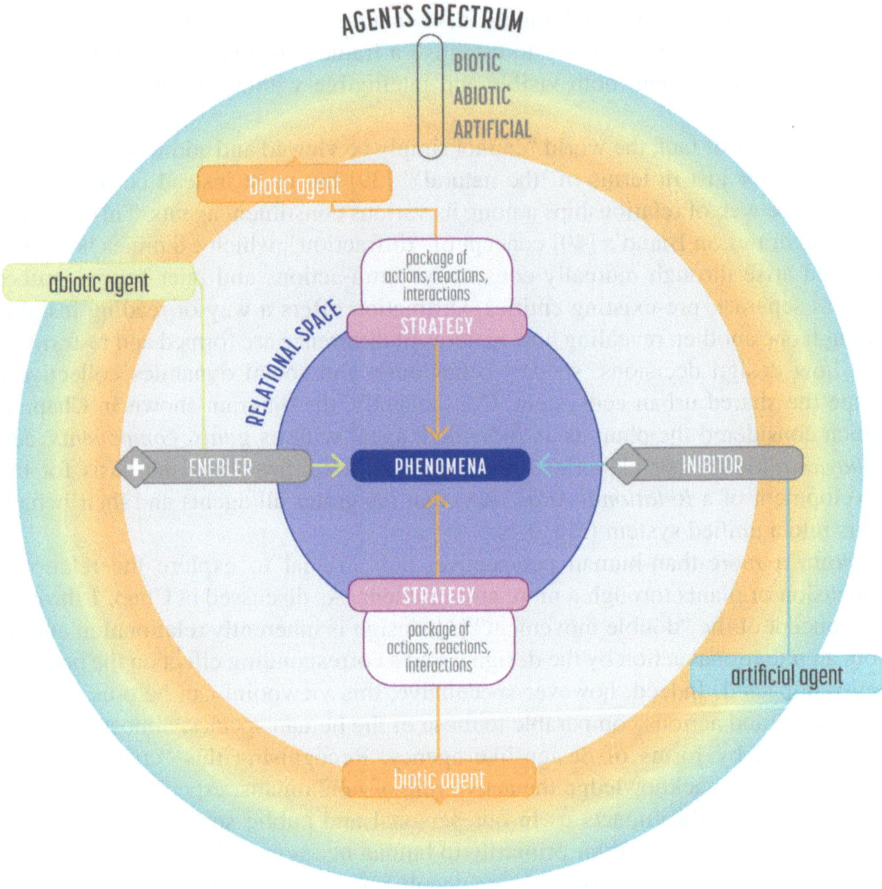

Fig. 3.1 The *Relational Atlas* composed by: (1) *Agents* (the ring outside with the *spectrum* between biotic, abiotic and artificial); (2) *Inibitors* and *enablers*; (3) *Strategies*; (4) *Phenomena*; (5) *Relational space*

phenomena, *relational space*, *enablers*, and *inhibitors*) are introduced here in a sequential manner, the *strategies* are discussed in detail in the following section.

Agents Agents are the driving force that comprises the *Atlas*, as they are the main characters acting within the system. To facilitate a more precise delineation of the agents fundamental to plant survival, these are structured within a ring incorporating an internal *spectrum* that differentiates between *biotic*, *abiotic*, and *artificial* agents, although the latter are not yet considered at this stage of RA development.

- ***Biotic factors*:** The biotic inner circle primarily comprises all living organisms. Using the biological classification system of kingdoms (Animalia, Plantae, Fungi, Protista, and Monera), the agents have been simplified for clarity into key categories: animals (further divided into mammals, insects, birds, and humans), plants, fungi, bacteria, and viruses. These agents, understood as *biotic factors*, are actively involved in various biological processes that influence plant health, development, and survival. Together, they form both visible and invisible networks of interaction, shaping the complex ecological relationships within which plants exist.

- ***Abiotic factors*:** The outer part of the ring comprises all those agents, such as light, air, soil, water, and temperature, that are essential to plant life. These *abiotic factors* play a vital role in the growth, survival and other *goals* of plants. They are just as critical as *biotic factors* in sustaining ecological balance and supporting the complex systems in which plants thrive. Although not grouped separately within the ring, the abiotic agents can be summarised as:
  - ***Light/Sunlight*:** light (and in particular sunlight) is used as an energy source for the production of sugars and starches, which are the plant's food. It is a form of electromagnetic radiation that is vital to photosynthesis in plants [41].
  - ***Air/Wind*:** air is the mixture of gases surrounding the Earth, and plants need to breathe and exchange gases thanks to photosynthesis [42]. As described by the philosopher Coccia [43], the air is the "cosmic fluid" generated by plants that connects all living beings in the atmosphere. Air, in the form of wind, also plays a crucial role in the pollination and seed dispersal of plants.
  - ***Soil*:** soil is the blend of organic and inorganic components covering the planet's surface [44]. It is the substrate in which plants grow and is vital for providing both nutrients and structural support.
  - ***Water/Rain*:** plants utilise water for various purposes, including nutrient transport, photosynthesis, and maintaining turgor. Therefore, its presence for plants is vital, and its absence causes plants to be unable to absorb essential nutrients for their well-being [45].
  - ***Temperature/Heat*:** temperature refers to the degree of heat present in a given environment and plays a fundamental role in shaping the metabolic rates, behaviour, growth, and survival of plants, both directly and indirectly.

Strategies The *actions, reactions, and interactions* discussed in the previous chapter provide specific insights into how plants behave in relation to both themselves and their surrounding environment. However, when biological knowledge is examined through a design lens, its specificity might become challenging to entirely integrate.

Indeed, in most cases, an agent does not rely on a single approach to achieve its objectives but instead engages in a sequence of actions. To better capture and communicate this complexity, I drew inspiration from Grime's [46] *plant strategy framework*, presenting these sequences as strategies employed by plants. Each strategy is supported by a portfolio of actions, reactions, and interactions, all of which are oriented toward the agents' *goals* described in the previous chapter. Moreover, each strategy is presented in relation to its degree of relationship with other hypothetical agents.

Phenomena Phenomena are the core of the *Atlas* and represent observable occurrences or processes that take place during the interaction between agents and the environment. The idea behind phenomena is that they represent something happening across all agents, without distinction of species, or even between living and non-living forms. It is often the manifestation of multiple interconnected factors, including ecological, biochemical, and relational. To develop the eight final phenomena, the material presented in Chap. 2 was systematised and subsequently compared with Grime's [46] plant strategy framework (Fig. 3.2).

The phenomena outlined so far are:

- ***Growth*:** the progressive development of an individual agent, marked by an increase in capacity, form, and internal complexity that enables maturation and establishment within a given environment.
- ***Adaptation*:** the modification of form, function, or behaviour in response to environmental change or external pressures, thereby supporting resilience and continuity over time.
- ***Assemblage*:** the formation of composite systems or communities through the integration of heterogeneous agents that interact, support one another, and coexist within dynamic relational configurations.
- ***Competition*:** the contest for limited resources, space, or favourable conditions, manifested through direct interactions (e.g., confrontation, exclusion) or indirect mechanisms (e.g., obstruction, resource depletion).
- ***Exploitation*:** an asymmetrical relational configuration in which one agent derives benefit from another, often to the latter's detriment, without necessarily threatening its continued existence.
- ***Expansion*:** the spatial extension of an agent or species through processes such as colonisation, dispersal, or reproduction, thereby increasing its ecological presence and influence.
- ***Regeneration*:** the restoration or reconstitution of energy, structure, or vital functions following disturbance, damage, or the completion of a life cycle, enabling renewed activity or growth.
- ***Decomposition*:** the breakdown of organic matter into simpler components that reintegrate into the ecological system, supporting subsequent biological processes and trophic relationships.

Relational space While phenomena are at the core of the system, the inner space where the relationships between agents are placed is called the *relational space*. It's the area of contention and negotiation, where agents put in place strategies to reach their objectives. The term *space* encompasses both the broader notion of a context in which processes and actions unfold and its more concrete physical dimension. The latter will be discussed in later chapters, particularly in relation to public spaces in cities. The former, on the other hand, suggests ideas of fluid spatiality and temporality that are typical of a more-than-human perspective [47]. As mentioned in Chap. 1, the issue of time is crucial: in scenarios of collaboration among agents of different species, the experience of time shifts from something relatively homoge-

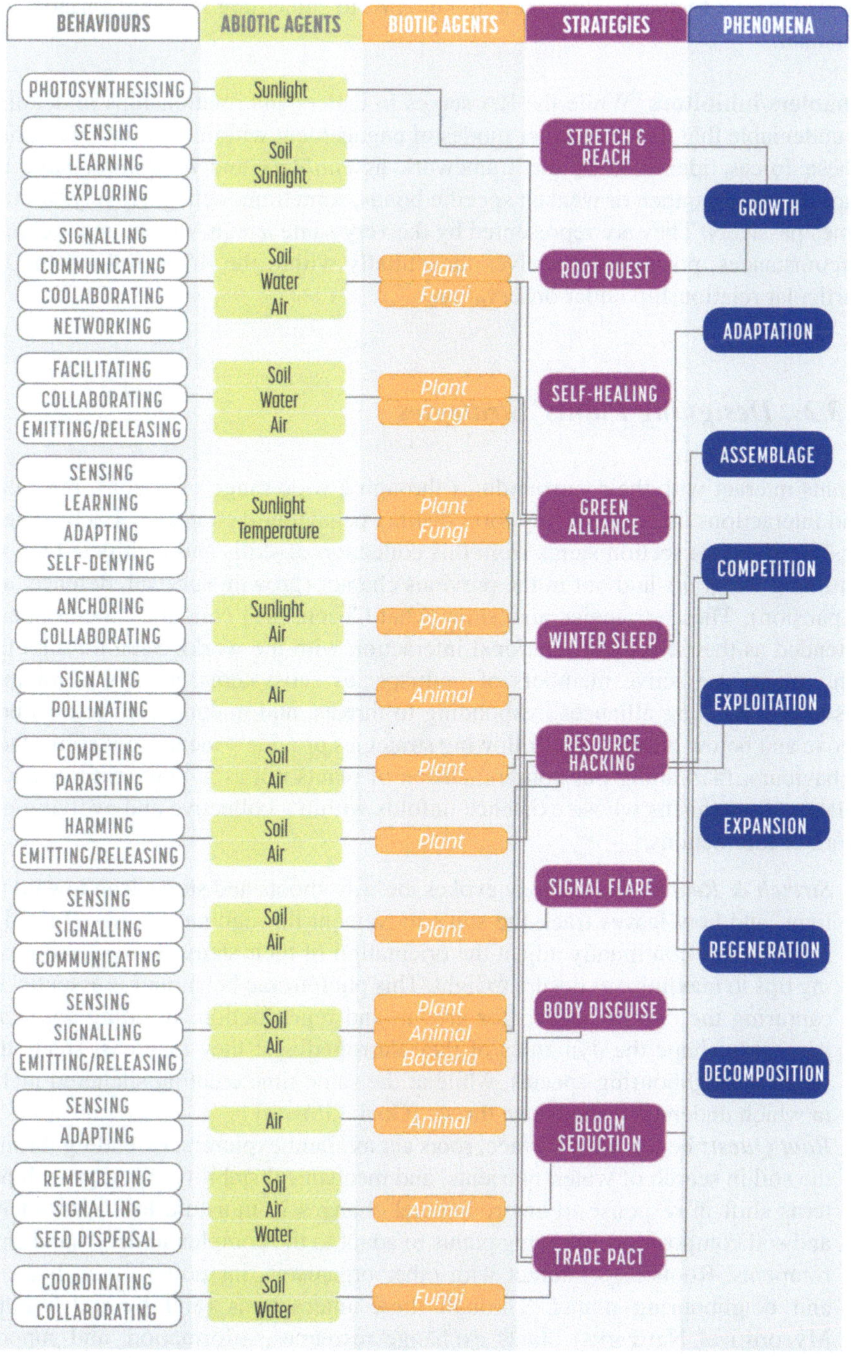

Fig. 3.2 The process of developing the *strategies* and *phenomena*. It is based on the material collected and elaborated in Chap. 2

neous (shared, to some extent, among humans) to something far more variable and personal.

Enablers/inhibitors While the RA serves to understand relationships in detail, it is undeniable that there are other modes of engagement within the relational space. These forces, identified in the framework as *enablers* and *inhibitors*, have the capacity to strengthen or weaken specific bonds, sometimes acting actively, at other times passively. They are represented by the very same agents which, under certain circumstances, position themselves tangentially within the *Atlas*, influencing the particular relationship under observation.

3.3.2 Designing Plants' Strategies

Plants interact with their surroundings through a wide range of actions, reactions, and interactions that collectively form distinct behavioural strategies. Each strategy described in this section stems from this collection of skills and is directed toward outlining the *goals* laid out in the previous chapter (growth, survival, defence, and expansion). These strategies also show what Coccia [43] calls the "social bond", intended as their constant, relational interaction with the world. Seen through this lens, plants are active members of multispecies ecosystems, always negotiating resources, forming alliances, responding to threats, and making connections both above and below ground. The following strategies provide a means to identify these behaviours, facilitating our comprehension of plants not as solitary entities but as relational organisms whose existence unfolds within a collective and evolving network of interactions.

- ***Stretch & Reach*:** this strategy evokes the way shoots and stems lean toward the light, and how leaves track the sun's movement throughout the day. Rooted in place, plants continually adjust the orientation of their stems, leaves, and growing tips to maximise exposure to light. This phototropic behaviour is essential for capturing the energy needed for growth and reproduction. In doing so, plants inevitably shape the dynamics of their surroundings: they may cast shade that restricts neighbouring species, while at the same time creating sheltered niches in which understory plants can thrive. [Goal: Growth]
- ***Root Quest*:** beneath the surface, roots act as silent explorers, extending through the soil in search of water, nutrients, and mechanical stability. Their growth patterns shift in response to environmental changes in moisture levels, obstacles, and soil composition, allowing plants to adapt to the complex underground environments. Roots also interact with other organisms, including fungi, bacteria, and neighbouring plants. Through these interactions (and by utilising the Mycorrhizal Networks) plants exchange resources, information, and support, creating a hidden web of relationships that is essential for the health of the wider community. [Goal: Growth]

- ***Self-Healing***: plants possess a remarkable ability to repair themselves. When damaged by herbivores, storms, or disease, they can seal wounds, isolate injuries, and redirect growth to restore lost structures. This regenerative ability reduces vulnerability to pathogens and limits opportunities for further attack, helping plants maintain resilience in unpredictable environments. [Goal: Growth]
- ***Green Alliance***: plants frequently form alliances, flexible, cooperative relationships with other species that strengthen their chances of survival. These partnerships can occur above ground, through volatile organic compounds, light signals, or chemical messengers, or below ground via expansive mycorrhizal networks. Within these alliances, plants share resources, warn one another of threats, and buffer external shocks, such as drought or extreme temperatures. Diverse plant communities exhibit greater resilience because each species contributes unique strengths, thereby creating a collective capacity to withstand disturbances. [Goal: Growth]
- ***Winter Sleep***: dormancy is a strategic withdrawal that allows plants to survive periods of cold, heat, or drought. By slowing their metabolism and conserving stored energy, they endure unfavourable conditions until the environment becomes hospitable again. During this pause, interactions with other organisms diminish, and survival depends largely on protective structures and internal reserves. [Goal: Growth]
- ***Resource Hacking***: plants often employ ingenious tactics to access resources in competitive environments. They may extend roots into nutrient-rich areas that others have not, tap into fungal networks for indirect nutrient uptake, or outcompete their neighbours through shading and rapid canopy expansion. Some species also employ chemical strategies, such as allelopathy, by releasing compounds that suppress their competitors. These actions influence community composition, shaping which species can survive or prosper in a given space. [Goal: Survival and Defence]
- ***Signal Flare***: plants can send chemical, visual, or electrical signals to protect themselves or alert others. They release volatile compounds to attract predators of herbivores, warn neighbouring plants to activate defences, or signal physiological stress. Such signalling influences the behaviour of insects, animals, and nearby plants, turning a single plant's distress into a coordinated defensive response. [Goal: Defence]
- ***Body Disguise***: some plants avoid danger by blending into their surroundings. Through camouflage (mimicry of shapes, colours, or textures), they reduce visibility to herbivores or deter unwanted attention. In some cases, mimicry serves the opposite purpose, attracting specific partners such as pollinators by resembling other species or environmental cues. [Goal: Defence]
- ***Bloom Seduction***: plants expand their lineage by attracting pollinators with bright colours, alluring scents, and nectar rewards. Through these sensory invitations, they enlist insects, birds, and mammals in the transfer of pollen, effectively outsourcing reproduction to mobile agents. This seductive strategy enables plants to spread across landscapes and ensures genetic diversity, relying on the movements and behaviours of other species to propagate. [Goal: Expansion]

- ***Trade Pact*:** many plants engage in reciprocal exchanges with other organisms, offering resources such as nectar, pollen, or shelter in return for pollination, seed dispersal, or protection. Above ground, chemical signals help attract the right agents, while below ground, mycorrhizal networks facilitate the exchange of nutrients that benefit both the plant and the fungus. These pacts often develop into long-term associations that stabilise ecosystems. They reflect a system of mutual benefit in which each participant contributes to, and depends on, the success of the other. [Goal: Survival and Expansion]

3.4 Discussing the *Relational Atlas*

The RA stems from both an urge to bring together data from other scientific fields and to support plants' agency in relation to others. In this sense, even if supported by existing and emerging knowledge, the way in which the RA is structured speculates the *intentionality* of plants. Drawing on the literature previously collected, whatever this intention might be, in the sense that we humans attribute to the term, is currently unknown. Therefore, in this part of the research, I decided to explore the concept of *intentionality* by filling the gaps in scientific knowledge and drawing inspiration from other worldviews that view nonhumans as active individuals. In this sense, I leaned more towards the knowledge derived from indigenous perspectives that recognise plants as *personas* to acknowledge their agency [24, 34, 35, 48].

The central premise of the RA is to position plants (and, when relevant, other agents) as primary actors in relation to their surroundings, thereby deliberately decentering the human perspective. It aims to reveal and articulate the full scope of their interactions with other life forms, making visible the depth of their radical interdependence [49] (Fig. 3.3).

Recognising this shared logic of action can serve as a crucial first step toward a broader shift in mindset that ultimately influences how design processes are conceived and practised. The *Atlas* should therefore be understood as a tool for examining relational processes between agents without filtering everything through a human-centric lens. It seeks to draw attention to the complexity of nonhuman life, which, although expressed in diverse forms, is grounded in the same fundamental principles of action, reaction, and interaction in response to vital processes and universal goals.

As for the research process, the RA functions as "a potential generator of knowledge" [50]. It emerged from an organisational need to make sense of the collected data and is ultimately transformed into a *designer's tool of inquiry*. In this context, the designer's perspective is not solely my own: it also symbolically incorporates the point of view of plants, however limited that representation may be. The RA is constructed around the needs, behaviours, and actions of plants, as interpreted through both scientific knowledge and indigenous perspectives, aiming to reflect their viewpoint as closely as possible.

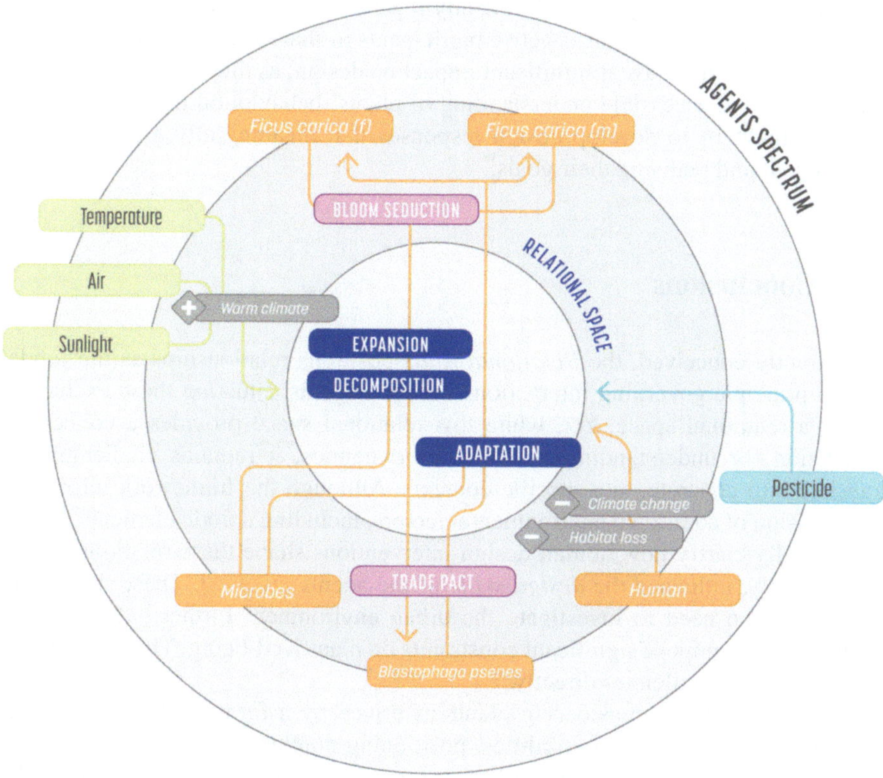

Fig. 3.3 The *Relational Atlas* showing the relationship between the common fig (*Ficus carica*) and the fig wasp (*Blastophaga psenes*)

At the same time, the plant *strategies* presented here can be understood as patterns of behaviour, adaptation, and interaction that enable plants to grow, survive, defend themselves, and, most importantly, coexist within their ecosystems. These strategies were developed by combining portfolios of *actions*, *reactions*, *interactions*, *goals*, and degrees of relationship. To prevent the anthropomorphisation of plants, I have attempted to characterise the strategies as objectively as possible. None of them is tied to the Cartesian notion of good or bad, but rather to the evolutionary thrust that every form of life seems to possess. In this respect, I also chose to highlight those *negative* (from a human perspective) strategies to go beyond a romantic vision of *nature* where everything is positive and collaborative.

In creating the *strategies*, specific attention was given to the notion of *social bonds* [38], which recognises that relationships with others are always central to growth and expansion. This concept proved especially valuable in considering multispecies dynamics, as it highlighted the different relational layers and the complex levels of interaction between plants and other agents. The strategies also illustrate how plants act not only as individuals but as communities that, much like humans, must engage with others to achieve their goals. This is a dimension of plant life that we often overlook, as plants are frequently regarded merely as decorative elements of urban contexts rather than as active participants in shared ecosystems. Therefore, these *strategies* may have a significant impact on design, as they offer a more transparent and more accessible understanding of plants' behavioural trajectories. This, in turn, allows us to develop design responses that meaningfully assist plants in carrying out and realising their goals.

3.5 Conclusions

As currently conceived, the *Relational Atlas* positions relationships as the fundamental principle governing interactions between agents, situating these exchanges within a relational space. Yet, while this relational space provides a conceptual foundation for understanding multispecies dynamics, it remains challenging to translate into concrete, site-specific contexts. Although the framework allows for the inclusion of additional environmental factors, including abiotic elements, it does not yet fully clarify how human design interventions shape these relationships or, consequently, influence the *design skills* of the agents involved. These limitations underscore the need to investigate the urban environment further, where design decisions often impose significant constraints on plant well-being. The next chapter addresses these challenges directly.

At the same time, considering plants as driven by a form of *design* opens up previously unimaginable possibilities, positioning nonhumans as legitimate stakeholders in decision-making processes. From a more-than-human lens, the *Atlas* should not be understood merely as a diagram of biological traits, but as a framework for rethinking design itself. It can support humans in moving beyond anthropocentric assumptions, recognising that other life forms possess their own needs, temporalities, and modes of problem-solving, and acknowledging plants as agents with goals, rather than passive resources. This perspective encourages the transition from extractive relationships to more reciprocal, mutually beneficial forms of coexistence.

Understanding plant *strategies* also lays the groundwork for enabling the cohabitation of multiple species. When we consider how plants behave, we can design urban and public spaces that work with their *strategies* rather than against them. In more-than-human and multispecies design processes, these *strategies* may ultimately serve as design prompts, rules of engagement, or participatory tools that foster more collaborative relationships between human and nonhuman communities.

References

1. Willis A-M (2006) Ontological designing. Des Philos Pap 4:69–92
2. (2016) No plant is an island. Nat Plants 2:16146. https://doi.org/10.1038/nplants.2016.146
3. Gorzelak MA, Asay AK, Pickles BJ, Simard SW (2015) Inter-plant communication through mycorrhizal networks mediates complex adaptive behaviour in plant communities. AoB Plants 7:plv050
4. Karban R (2008) Plant behaviour and communication. Ecol Lett 11:727–739
5. Dudareva N, Klempien A, Muhlemann JK, Kaplan I (2013) Biosynthesis, function and metabolic engineering of plant volatile organic compounds. New Phytol 198:16–32
6. Hedrich R, Salvador-Recatalà V, Dreyer I (2016) Electrical wiring and long-distance plant communication. Trends Plant Sci 21:376–387
7. Penuelas J, Llusià J (2001) The complexity of factors driving volatile organic compound emissions by plants. Biol Plant 44:481–487
8. Runyon JB, Mescher MC, De Moraes CM (2006) Volatile chemical cues guide host location and host selection by parasitic plants. Science 313:1964–1967
9. Zebelo SA, Maffei ME (2012) Signal transduction in plant–insect interactions: from membrane potential variations to metabolomics. In: Plant electrophysiology. Springer, pp 143–172
10. Delory BM, Delaplace P, Fauconnier M-L, Du Jardin P (2016) Root-emitted volatile organic compounds: can they mediate belowground plant-plant interactions? Plant Soil 402:1–26
11. Falik O, Mordoch Y, Quansah L et al (2011) Rumor has it…: relay communication of stress cues in plants. PLoS One 6:e23625
12. Kegge W, Pierik R (2010) Biogenic volatile organic compounds and plant competition. Trends Plant Sci 15:126–132
13. Lawrence JL, Wiebe SM (2018) Biopolitical disaster. Taylor & Francis
14. Roudavski S (2020) Multispecies cohabitation and future design. In Boess S, Cheung M, Cain R (eds) Synergy - DRS International Conference 2020. https://doi.org/10.21606/drs.2020.402
15. Von Essen E, Allen MP (2017) Solidarity between human and nonhuman animals: representing animal voices in policy deliberations. Environ Commun 11:641–653
16. Centenaro G, Hudek C, Zanella A, Crivellaro A (2018) Root-soil physical and biotic interactions with a focus on tree root systems: a review. Appl Soil Ecol 123:318–327
17. Köhler M (2008) Green facades—a view back and some visions. Urban Ecosyst 11:423–436. https://doi.org/10.1007/s11252-008-0063-x
18. Raven JA, Edwards D (2001) Roots: evolutionary origins and biogeochemical significance. J Exp Bot 52:381–401. https://doi.org/10.1093/jexbot/52.suppl_1.381
19. Trewavas A (2017) The foundations of plant intelligence. Interface Focus 7:20160098
20. Macfarlane R (2016) The secrets of the Wood Wide Web. The New Yorker
21. Viola A, Mancuso S (2015) Verde brillante. Sensibilità e intelligenza del mondo vegetale. Giunti Editore, Firenze
22. Ciszak M, Comparini D, Mazzolai B et al (2012) Swarming behavior in plant roots. PLoS One 7:e29759
23. Vacchiano G (2019) La resilienza del bosco: storie di foreste che cambiano il pianeta. Mondadori, Milano
24. Hall M (2011) Plants as persons: a philosophical botany. Suny Press
25. DelSesto M (2020) People–plant interactions and the ecological self. Plants People Planet 2:201–211
26. Jose SB, Wu C, Kamoun S (2019) Overcoming plant blindness in science, education, and society. Plants People Planet 1:169–172
27. Parsley KM (2020) Plant awareness disparity: a case for renaming plant blindness. Plants People Planet 2:598–601
28. Kulick D (2017) Human-animal communication. Annu Rev Anthropol 46:357–378
29. Marder M (2013) Plant-thinking: a philosophy of vegetal life. Columbia University Press

30. Laist R (2013) Plants and literature: Essays in critical plant studies. Rodopi B. V., Amsterdam - New York
31. Rice L (2018) Nonhumans in participatory design. CoDesign 14:238–257
32. Abram D (1996) The spell of the sensuous: perception and language in a more-than-human world, 2017th edn. Vintage, New York
33. Harvey G (2014) The handbook of contemporary animism. Routledge
34. Abbott S (2021) Approaching nonhuman ontologies: trees, communication, and qualitative inquiry. Qual Inquiry 27:1059–1071
35. Brennan A, Lo N (2002) Environmental ethics. In: Zalta EN, Nodelman U (eds) The Stanford Encyclopedia of Philosophy. The Metaphysics Research Lab, Stanford
36. Mathews F (2011) Towards a deeper philosophy of biomimicry. Organ Environ 24:364–387
37. Viveiros de Castro E (2012) Cosmological perspectivism in Amazonia and elsewhere. Masterclass Series 1, HAU Network of Ethnographic Theory, Manchester
38. Coccia E (2021) Métamorphoses. Wiley
39. Fry T (2009) Design futuring. University of New South Wales Press, Sydney, pp 71–77
40. Barad K (2014) Diffracting diffraction: cutting together-apart. Parallax 20(3):168–187. https://doi.org/10.1080/13534645.2014.927623
41. Niinemets U (2007) Photosynthesis and resource distribution through plant canopies. Plant Cell Environ 30:1052–1071
42. Bell JNB, Treshow M (2002) Air pollution and plant life. Wiley
43. Coccia E (2018) La vita delle piante: metafisica della mescolanza. Società editrice il Mulino
44. El-Ramady HR, Alshaal TA, Amer M et al (2014) Soil quality and plant nutrition. In: Sustainable agriculture reviews, vol 14. Springer, pp 345–447
45. Chaves MM, Pereira JS, Maroco J et al (2002) How plants cope with water stress in the field? Photosynthesis and growth. Ann Bot 89:907–916
46. Grime JP (1977) Evidence for the existence of three primary strategies in plants and its relevance to ecological and evolutionary theory. Am Nat 111:1169–1194
47. Bessai R, Bendor R, Balkenende R (2024) Designing with more-than-human temporalities. In: Gray C, Ciliotta Chehade E, Hekkert P, Forlano L, Ciuccarelli P, Lloyd P (eds) DRS2024: Boston, 23–28 June, Boston, USA. https://doi.org/10.21606/drs.2024.438
48. Gibson D (2018) Towards plant-centred methodologies in anthropology. Anthropol South Afr 41:92–103
49. Escobar A (2018) Designs for the pluriverse: radical interdependence, autonomy, and the making of worlds. Duke University Press
50. Stappers PJ, Giaccardi E (2017) Research through design. In: The encyclopedia of human-computer interaction. The Interaction Design Foundation, pp 1–94

Chapter 4
Rethinking Public Space for Situated Multispecies Collectives

Abstract While vegetation in public spaces is widely acknowledged to improve climate comfort, air quality, biodiversity, and social well-being, most of these considerations are still based on anthropocentric assumptions that prioritise human benefits over plant health. Urban development often disrupts existing ecosystems, removing mature vegetation and introducing new plants into environments characterised by soil compaction, pollution, resource scarcity, and extreme weather. Building on the *Relational Atlas* presented in the previous chapter, this chapter investigates how urban planning has historically overlooked nonhuman needs and considers whether cities can be reimagined as multispecies collectives. Through a collection of plant *stressors* and field observations, it argues for a holistic, ecosystemic approach to planning that recognises plant agency, addresses the cumulative pressures of urban life, and positions nonhuman wellbeing as critical to just and regenerative urban futures.

4.1 Introduction

Integrating plants into the design of public spaces contributes significantly to creating more livable and sustainable cities. Public spaces, such as parks, squares, and sidewalks, often incorporate trees and other vegetation as key landscape features [1]. These integrations not only enhance the visual appeal of cities but also provide shade, mitigate heat, improve air quality, and support urban biodiversity by offering habitats for *wildlife* [2]. Beyond their environmental functions, plants also play a vital role in shaping a sense of place and fostering social interaction and community gathering [3]. Many communities around the world have adopted strategies to integrate ecological processes into cities through locally tailored, resource-efficient, and systemic design practices. While often complex to implement at scale, such solutions are closely intertwined with the overall well-being of urban populations, underscoring the essential role of plants and other nonhumans in the health and resilience of contemporary cities.

F. Vergani, *Reframing Urban Ecologies*, SpringerBriefs in Applied Sciences and Technology, https://doi.org/10.1007/978-3-032-18924-0_4

However, from a more-than-human perspective, it remains challenging to assess the extent to which these urban *greening* strategies genuinely address the well-being of nonhumans. Most notably, such solutions rarely transcend a fundamentally anthropocentric worldview. Urban development typically begins with large-scale removal of vegetation, resulting in the destruction of habitats and the loss of mature trees—many of which have taken decades or even centuries to grow [4]. At the same time, once integrated into the urban fabric, new plants face a host of *stressors*: limited access to sunlight, water, and nutrients due to surrounding infrastructure; soil compaction; air pollution; and exposure to extreme weather events, all of which compromise their health and longevity [4–8].

This chapter opens with these reflections, which, among other factors, first guided me toward exploring the field of more-than-human design. To fully grasp how plants act within the relational space outlined in the previous chapter through the *Relational Atlas*, it is essential to examine how urban planning has historically addressed—or neglected—the presence and needs of nonhuman life. Tracing this trajectory offers necessary context for understanding the challenges plants currently face in urban environments. The more-than-human shift invites us to ask whether it is possible to reimagine urban communities as *situated multispecies collectives*, in which nonhuman agents are not only acknowledged but embraced as living and potentially collaborative participants in shaping shared space.

4.2 Plants Living in Cities

Setting aside the vast diversity of the plant kingdom, we can broadly categorise plants into two simplified groups: (1) those that grow or are cultivated in *naturalistic* contexts such as forests, and (2) those that emerge or are cultivated within urban environments. Despite being a prominent feature of the urban landscape, plants have historically been overlooked in descriptions and conceptualisations of the city [9]. Urban settlements are typically understood as human-centric spaces, *artificial* environments designed by and for people. Since the establishment of the Greek *polis*, plants and other nonhumans have been systematically excluded from the political and spatial discourse of the city [10]. In Ancient Rome, for instance, vegetation was primarily relegated to sacred groves, distant from the civic heart of the city [11]. This exclusion—an *original sin* of Western culture when viewed from a more-than-human perspective—has contributed to the entrenched dualism that frames everything beyond the city walls as *natural* or *wild* [12]. The common misconception that cities belong solely to humans, while forests are the domain of *nature*, persists in both urban theory and everyday thinking. Yet, many ecosystems we regard as natural (such as forests) have in fact been shaped by long-forgotten human interventions. These actions have altered their biological composition, manipulating plant shape or introducing non-native species [13, 14]. Over time, such forms of control and regulation have led to the fragmentation of ecosystems, resulting in residual zones and the emergence of new types of boundaries [15, 16].

Gilles Clément [17] addresses this phenomenon in his *Manifesto of the Third Landscape*, defining these residual spaces as territories for the many species that cannot find a place elsewhere. Known as *friches* in Clément's terms (or *ecotones* in ecological science [14]) these interstitial areas often host a remarkable concentration of diverse biological life forms within small, overlooked patches of the urban fabric. Urban environments, constantly reshaped by human activity, can therefore be reimagined as sites of "mixture" [18] or "cross-fertilisation" [17], where humans and nonhumans coexist, act, and collaboratively design the spaces they inhabit.

The growing body of knowledge emerging from more-than-human studies calls for a fundamental rethinking of what cities are and who they are for, urging us to reconsider the role of plants, particularly trees, in a more inclusive and meaningful way. Concepts such as the "multispecies city" [19] and the "more-than-human city" [20] challenge traditional notions of urban space by proposing models that actively consider nonhumans in planning and design processes. However, implementing these ideas remains a significant challenge, as most urban environments are still conceived through an anthropocentric lens. As Cavender and Donnelly [5] emphasise, citing Jim [7], urban trees "are faced with a complex array of stressors that often prevent them from realising their biological potential in terms of tree form, health, and life span." In densely built landscapes, plants must contend with numerous pressures, from limited light, compacted soils, and pollution to the physical constraints imposed by buildings and underground infrastructure [5].

From this perspective, cities can be seen as hostile environments for plants. Addressing this condition requires a shift in urban planning and design toward a more-than-human, ecosystemic approach that foregrounds the interdependence of all life forms and actively pursues innovative strategies to support the well-being of both humans and nonhumans. Such a transformation not only enhances biodiversity and ecological resilience but also redefines the city as a shared, collaborative space for all beings.

4.2.1 Strategies for "Greening" Cities

Ecosystems across the planet have been profoundly altered by processes of anthropisation, with cities emerging as particularly toxic for public health and ecological resilience [21–27]. As outlined in the United Nations Sustainable Development Goal 11—*Sustainable Cities and Communities*—, cities are increasingly burdened by rapid population growth, climate change, and rising levels of air pollution. In response to these challenges, the past few decades have seen the emergence of design practices focused on the concept of *green* urbanism. These efforts aim to reintroduce *natural features* into city planning as a means of mitigating environmental harm and improving livability. The origins of urban greening date back to the late nineteenth and early twentieth centuries, when architects and planners began to explore new models of urban living that could address public health crises and the unhygienic conditions of rapidly industrialising cities [28]. The need for greenery became even more urgent in the wake of the Second Industrial Revolution, as cities struggled to cope with the environmental consequences of increased energy consumption, particularly from fossil fuels [29].

Beginning in the late nineteenth century, both Europe and the United States began experimenting with urban design strategies that aimed to integrate residential, industrial, and agricultural zones within cities. Concerned with the overcrowding and decline of urban areas, urban planner Ebenezer Howard (1965) introduced the *Garden City Movement*, a visionary model based on a concentric layout featuring open spaces, public parks, and six radial boulevards designed to create a harmonious environment for living and working. Although this model was only fully realised in a few British cities, it went on to inspire similar urban planning approaches across various global contexts [30].

However, it wasn't until the late twentieth century that the concept of *green* cities became embedded in a broader urban vision, emphasising not only the environmental and health benefits of green spaces but also their role in enhancing urban livability, attracting visitors, and promoting sustainability [31, 32]. The 1992 Earth Summit marked a pivotal moment in global environmental discourse, serving as one of the first international platforms where representatives from nations around the world came together to assess ecological challenges and establish a shared agenda for sustainable development. From that point forward, the theme of *sustainability* became central to design practices across disciplines [29], prompting a shift in how plants were perceived in urban contexts. Rather than being seen solely as aesthetic features, plants began to be recognised as strategic tools for addressing pressing global issues such as climate change and biodiversity loss. The integration of vegetation into cities offers a wide range of environmental benefits, including improved air quality, temperature regulation, support for biodiversity, and enhanced human well-being [5, 33–35]. As Czaja et al. [4] notes, "in recent years, the function of urban trees has changed [since they] are no longer only ornamental or a source of raw materials, as they also provide measurable environmental benefits." This shift reflects a broader recognition of the vital role plants (particularly trees) play in urban ecosystems, especially when they are planted in communities to reassemble urban forests. These initiatives aim to recreate the complexity and resilience of forest systems within urban environments [5, 35–38]. As Turner-Skoff and Cavender [35] clearly outline, the benefits of trees can be categorised into five key domains: (a) health and social well-being; (b) cognitive development and education; (c) economy and resource efficiency; (d) climate change mitigation and habitat creation; (e) green infrastructure.

From this perspective, biodiversity-driven approaches and strategies, such as Nature-Based Solutions (NBS) [39–45], are increasingly *blooming* within urban environments as responses to the urgent challenges of climate change, urban degradation, and ageing infrastructure. The European Union defines NBS as:

> "solutions that are inspired and supported by nature, which are cost-effective, simultaneously provide environmental, social and economic benefits and help build resilience. Such solutions bring more and more diverse nature and natural features and processes into cities, landscapes and seascapes, through locally adapted, resource-efficient and systemic interventions".

NBS are being integrated into urban contexts through a variety of approaches, ranging from regenerative agricultural practices, such as *permaculture* [46, 47], to large-scale systems like *green infrastructure* [36, 44]. Over the past 50 years, permaculture has played a pivotal role in fostering inclusive, biodiverse landscapes that support both ecological integrity and human well-being. Developed in the 1970s by Bill Mollison and David Holmgren [47], permaculture is grounded in a set of design principles that aim to create ecosystems capable of sustaining self-sufficient human settlements. These principles encompass everything from polyculture systems, featuring perennial trees, shrubs, herbaceous plants, fungi, and roots, to the architectural and energy management strategies needed to support them [46]. More than a technical system, permaculture cultivates a profound sense of connection and responsibility between human and nonhuman members of a shared ecosystem, encouraging a culture of care for the environment and its complex entanglements [48].

The values and principles of permaculture have been widely adopted in the design of community gardens, where the synergy between human and nonhuman life is more acknowledged. These spaces exemplify a model of coexistence where ecological and social functions reinforce one another, promoting environmental stewardship and community resilience [49–51]. Community gardens offer a highly democratic model of design and management, where plants, insects, and other small animals play essential roles in maintaining ecological balance [50]. The effectiveness of NBS in such contexts is also evident in how these spaces often embrace minimal human intervention, allowing nonhumans to guide the process of spontaneous *rewilding* and regeneration.

In parallel, urban planners and designers are increasingly integrating *green infrastructures* into cityscapes to provide essential ecosystem services. These interventions include *green corridors* [36, 44, 52, 53], *green* or *living roofs* [54–58], and *green walls* [54, 59–61], as well as other strategies to enhance water availability and quality. Green infrastructures are networks of (semi-)*natural* areas which are protected and enhanced to deliver ecosystem services, while also benefiting biodiversity and society more widely. In this framework, plants and trees serve as core features, alongside water, in shaping new ecosystems that support biodiversity and provide a wide range of environmental, social, and economic benefits.

Today, designers, urban planners, and policymakers are making coordinated efforts to integrate NBS into the planning guidelines of municipal administrations, promoting ecological sustainability through the reintegration of plants and their intricate web of nonhuman relationships. In this context, *sustainable smart cities* have emerged as key players, employing advanced technologies to monitor and manage the health of urban ecosystems while enhancing human quality of life [62–65]. Despite this progress, these emerging design approaches still fall short in three significant ways. First, NBS projects often struggle to create meaningful impact at the city scale. Many existing initiatives remain fragmented, highly localised, and disconnected from one another, limiting their systemic influence [66]. Second, urban governance plays only a marginal role in fostering urban biodiversity due to limited funding, scarce resources, and poor coordination among relevant stakeholders and agencies [45, 67]. As Frantzeskaki [42] notes, "there is yet limited research on the ways the knowledge on NBS can advance urban policy and planning to enable their mainstreaming." Third, although these strategies contribute meaningfully to urban biodiversity, they remain predominantly anthropocentric, geared toward enhancing human health and well-being [35], rather than genuinely reimagining the city as a shared habitat for multiple species [68–72].

4.2.2 *Different Scales of the Same Topic*

Over the past 15 years, many local governments and municipal authorities have adopted NBS, promoting tree planting and conservation programs, and introducing regulations to prevent the removal or harm of urban plants. Over time, cities have built global networks, collaborations, and partnerships to promote sustainable and resilient urban development. These initiatives connect urban planners, policymakers, and stakeholders to exchange knowledge, skills, and best practices in the pursuit of innovative and inclusive solutions. Aligned with the United Nations' 2030 Agenda, these networks work collectively to address the complex demands of global urbanisation. Among the most prominent are the Global Covenant of Mayors for Climate and Energy, C40 Cities, the Resilient Cities Network, and ICLEI (Local Governments for Sustainability). These organisations collaborate with partners to integrate sustainable practices across various domains, including rethinking how cities engage with ecosystems and the nonhumans within them.

Although these networks often advocate for healthier, more livable cities by reclaiming space for people and other living agents, improving air quality, and fostering vibrant local communities, they still fall short in recognising the cognitive capacities or agency of plants. As articulated in Chap. 2, there is little to no evidence of a shift in sensitivity toward understanding plants as sentient or intelligent beings within the frameworks of these initiatives. One partial exception is found in an official report from ICLEI, developed in collaboration with the IUCN and The Nature Conservancy through their *CitiesWithNature* platform, a global initiative encouraging cities to commit to more sustainable urban environments by enhancing the role of *nature*. In a 2021 publication, the program introduced a promising shift in perspective by spotlighting the nonhuman world and promoting a multispecies approach to urban development. The report emphasised the need to integrate the interests and needs of plants, animals, and broader biodiversity into urban planning and policymaking processes [45]. It also highlighted the importance of involving citizens, institutions, and businesses in caring for other species.

However, despite this forward-thinking rhetoric, nonhumans are quite always still framed mainly in instrumental terms. Their value is predominantly justified by the benefits they offer to human well-being, rather than being recognised as autonomous stakeholders in their own right. As such, even within more progressive initiatives, the narrative remains firmly rooted in anthropocentrism.

4.2.3 Plants in the Everyday Life of Human Communities

Given the wide range of design and urban planning strategies aimed at integrating plants and other biological life into cities, it becomes essential to reframe the urban perspective through a more localised lens. This means exploring how plants, beyond being resources for urban sustainability, can be meaningfully integrated into the discourse of community at a smaller, more situated scale. Even within limited portions of the city, neighbourhoods function as complex ecosystems composed of diverse life forms interacting daily within specific spatial and infrastructural frameworks [73]. These urban *patches* are dynamic zones shaped by agents, flows, and interactions, and often develop distinct identities rooted in historical and cultural processes [74, 75].

In specific neighbourhoods, this dynamic takes on greater significance, often driven by the proactive engagement of individuals—whether members of the so-called "creative class" [76] or "professionals of the everyday" [77]—who tackle local challenges and contribute ideas and perspectives that shape the community. This process of "active involvement" typically stems from a socially and culturally equipped segment of the population. However, it is also common for many neighbourhood *voices* to go unheard [73]. These overlooked voices include those "under the radar" [78], such as groups with a "low degree of social resilience" [79], including migrants, older people, children, and individuals with physical or cognitive disabilities. Critically, they also include nonhumans whose presence is often entirely excluded from participatory processes and urban discourse.

From a more-than-human perspective, plants should be regarded as members of neighbourhood communities, contributing to the creation of environments that are both pleasant and livable for all. Beyond the many health benefits they provide, plants are essential components of urban life, engaging with humans in multiple ways—whether these interactions are mutual or asymmetrical, and whether they are consciously acknowledged or not. As detailed in Chap. 3, we interact with plants daily [80, 81]. They provide visual cues about weather, climate stress, and seasonal changes, while physically offering food, shade, and relief from heat during warmer months. As such, plants play a pivotal role in shaping the human experience [82], influencing our social, physical, psychological, cognitive, environmental, and spiritual well-being [83].

Given these premises, there is room to consider plants as citizens who play an essential part in the urban ecology, even if people still struggle to see them as active agents and actual stakeholders. On the one hand, there is a shared commitment to preserving urban greenery as a public good; on the other, a persistent reluctance remains to move beyond anthropocentric perspectives and to fully recognise plants as more-than-human participants in the city. Elinor Ostrom's [84] theory of the *commons* provides valuable insights into how communities can collectively manage shared, vulnerable resources (such as forests, fisheries, air, and water) through cooperative practices that operate outside of market or government control [85, 86]. While this framework effectively highlights the role of citizens in stewarding those resources, the more-than-human paradigm invites a further shift in perspective. Instead of considering plants as resources to be extracted and utilised, it calls for their recognition as agents to be integrated as members of the urban community.

4.3 Grounding the *Atlas*: The Relational Space in Real Life

In the previous chapter, the *Relational Atlas* (RA) was framed solely in terms of the relational dimension. While the RA offers a situated perspective on relationships, each context is shaped by its own distinct logics and dynamics, making them far from universal.

Within the more-than-human framework adopted here, the most compelling aspect lies in examining relational dynamics through the lens of *situatedness*. Articulating on the concept of "situated knowledge" by Haraway [87], the term here refers to the idea that phenomena are deeply embedded within specific contexts, influenced by experiences, the environment, and the particular task at hand.

Yet, the spatial perspective inevitably shifts depending on the scale under consideration. For this reason, the *Atlas* requires further refinement, which can be achieved by introducing an additional spatial dimension alongside the relational one. This added dimension should be understood as a lens applied to the framework, allowing for a reassessment of key characteristics such as biotic and abiotic agents. This spatial lens is structured around three focal perspectives: *macro*, *meso*, and *micro* (Fig. 4.1).

By *macro*, I refer to large-scale ecological sites such as nature reserves or protected landscapes, where ecological processes unfold across extensive areas and involve complex multispecies dynamics; by *meso*, I mean urban parks and comparable green infrastructures that mediate between human activities and environmental functions at the neighborhood or city level; and by *micro*, I point to smaller, more intimate spaces such as flowerbeds, courtyards, or domestic gardens, where interactions are localized yet still carry significant ecological, cultural, and affective value. Each contextual scale brings with it different actors, policies, tools, methods, and financial resources. Within this framework, the RA operates at the micro scale, highlighting how design also, and above all, emerges from a relational and caring process that acknowledges the agency of all plants.

Alongside this conceptual expansion of the *Atlas*, the chapter presents a set of field observations that revealed some of the limitations plants perceived in cities. What emerges from this ethnographic work, echoing insights from the literature, is the presence of multiple stress factors within urban public spaces, which often restrict and undermine many of the plants' objectives. Moreover, the presence of these *stressors* further sharpens the role that inhibitors play within the RA.

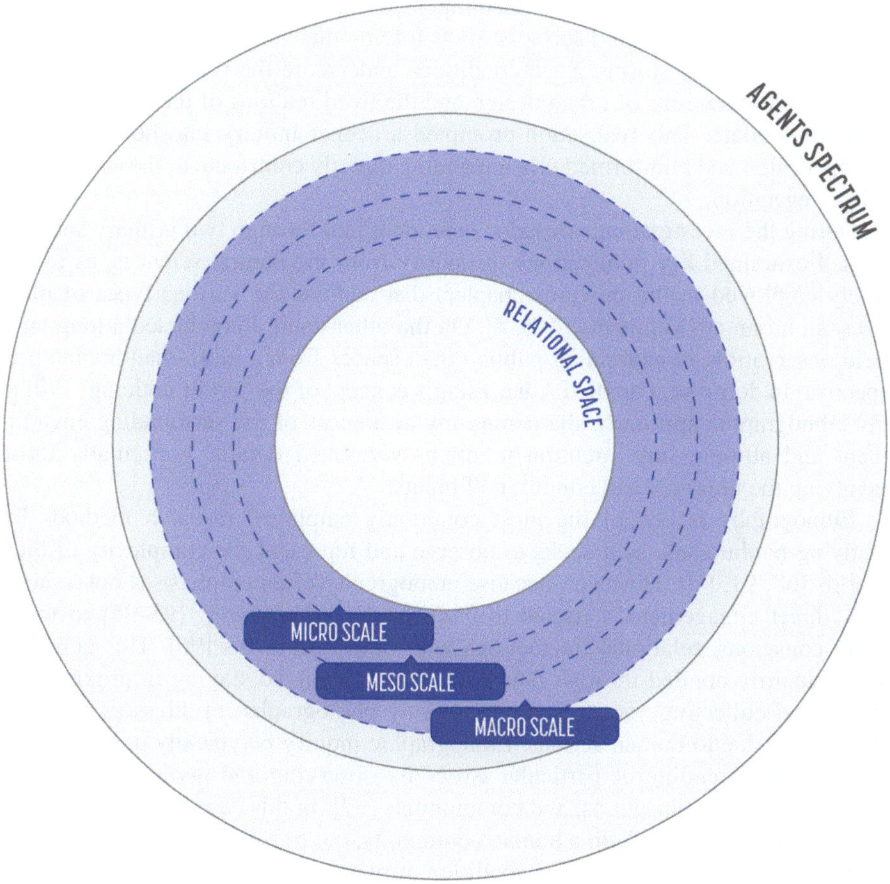

Fig. 4.1 The three different scales of the relational space embedded into the *Relational Atlas*: *macro*, *meso*, and *micro*

4.3.1 Searching for Stressors

Starting from the previously articulated premise that public urban spaces are predominantly designed through an anthropocentric lens, it becomes evident that they are almost exclusively shaped for the benefit of human beings, adhering to rules and standards that prioritise human needs. Within this framework, plants are typically regarded as aesthetic or functional features that support human well-being [35, 88, 89], rather than as agents with their own needs, behaviours, and ecological roles.

One of the earliest motivations behind my ongoing investigation was the precarious condition of plants in urban environments. It is not difficult to observe, daily, the challenges and mismanagement plants face in public spaces. As previously discussed, plants contend with a wide range of *stressors* that frequently prevent them from reaching their full biological potential in terms of form, health, and longevity.

In urban environments, it is not uncommon to encounter trees scorched by heat, trunks constrained by paved surfaces, roots forcing their way through concrete, and branches broken by storms. Such conditions underscore the profound detachment between the intentions of urban design and the lived realities of the plants it seeks to accommodate. This realisation prompted a deeper inquiry into how a human-centric design and uninformed urban planning directly contribute to the suffering of urban vegetation.

During the investigation, *stressors* were identified through two primary sources. First, I examined key publications (primarily from the natural sciences, as extensively employed in the previous chapter) that address the various types of plant stress in urban environments [4–6, 8]. On the other hand, I conducted a long-term field observation, re-examining public urban spaces from a more-than-human perspective. In doing so, I applied Anna Tsing's concept of the "art of noticing" [90] to my ethnographic approach, sharpening my awareness of the surrounding environment and attuning my attention to often-overlooked details, particularly those involving the presence and condition of plants.

Ethnography is one of the most commonly employed research methods for studying nonhumans, as it seeks to observe and represent the complexity of their worlds [68, 91, 92]. However, because ethnography often emphasises observation over direct engagement, I turned to *multispecies ethnography* [93–95] to foster more conscious, relational interactions with other life forms [96]. This approach was primarily applied through fieldwork, which involved staying in proximity to plants and collecting visual evidence (mainly photographs) of stressors affecting urban trees due to human actions. Ethnographic inquiry is typically used to gain a holistic understanding of particular issues by observing and participating within groups, teams, organisations, and communities [97]. In this case, however, the goal was not to collect data from a human community, but rather to document and interpret visual information about the condition of plants within cities.

Given the inherent challenge of engaging with nonhumans, I extended the ethnographic method beyond observation and the act of noticing [98] by integrating additional more-than-human techniques, such as *attuning* to [68, 99], that is, harmonising with the context and attempting to *listen* to the needs of plants. Attuning follows proper more-than-human methods of *decentering* [100–102], which involves shifting perspective away from a human-centred worldview and, in this specific case, adopting that of plants. In this way, multispecies ethnography resonates with Indigenous ways of knowing, which regard plants as sentient and deserving of respect [68, 103]. Since plants suffering in urban environments cannot "speak" in human terms, this visual documentation aims to bring their experiences into focus, placing vegetal life at the centre and exposing the harm inflicted on them by human design decisions and municipal policies.

To narrow down the inquiry, I decided to limit the research to medium and tall trees characterised by a dominant trunk and a well-defined crown shape [104]. These trees play a vital role in stabilising soil to prevent erosion, regulating local climate conditions, and contributing to atmospheric balance by absorbing carbon dioxide and producing oxygen [105].

4.3.2 Stressors Identification

As previously described, a non-structured multispecies ethnography marked the first turning point that guided me toward the field of more-than-human design. Over the past 5 years, walking through urban environments across various parts of the world has allowed me to compile a collection of photographs that capture the condition of plants (Figs. 4.2 and 4.3). These early field observations served as the initial spark, deepening my awareness of the challenges faced by nonhuman life in cities. However, to move from impressions to a more structured and scientifically grounded understanding, I turned to key academic publications [4–6, 8], which reinforced and refined my knowledge of plant *stressors* in urban contexts. The photographs taken during fieldwork were then reviewed, sorted, and organised into thematic clusters. Grounded in the belief that cities are shared habitats populated by a diversity of agents (not solely humans) and drawing on Fry's [106] concept of "ontological relationality"—which emphasises the deep interdependence between agents and their environments—, I grouped the identified stressors into two main categories:

- **Environmental-induced stressors:** These stressors are pressures that stem from *natural* conditions (both *abiotic* and *biotic*) that can hinder a plant's ability to grow, reproduce, or survive. These include climatic challenges such as drought, flooding, frost, and heat waves, as well as soil-related issues like salinity, nutrient shortages, or contamination. Biotic pressures, such as competition with neighbouring species, damage from herbivores, or attacks by pathogens, also belong to this category. Although such *stressors* are part of ordinary ecological dynamics, their frequency and severity have increased under climate change, amplifying their impact on plant life in both *natural* and urban environments.
  - Climate change;
  - Land-cover change.
- **Human-induced stressors:** Human-induced stressors arise from activities and decisions that directly or indirectly compromise plant health. Urbanisation and land-use change, pollution (from particulates, heavy metals, or chemical runoff), trampling, and mechanical damage in public spaces all place stress on plants. Maintenance practices can exacerbate these problems: excessive pruning, pesticide use, or soil compaction often weaken rather than support plant vitality. Some effects are less visible, such as the creation of heat islands or the reduction of biodiversity when ornamental species are favoured over native ones. Together, these pressures demonstrate how human actions alter plant life in ways that often compromise resilience and ecological functioning.
  Above-Ground
  - Lack of direct sunlight;
  - Artificial light;
  - Restricted space for crown growth;
  - Poor pruning and maintenance;
  - Pollution from vehicles, infrastructures, and runoff;

Fig. 4.2 Plants *stressors* identified while walking in Milan (Italy)

Fig. 4.3 Plant *stressors* identified while walking in New York City (USA)

  - Air temperature;
  - Increased risk from pests and disease;
  - Pedestrians and vehicles' passage.
- Below-Ground
  - Restricted root space;
  - Poor soil composition and structure;
  - Soil compaction;
  - Poor surface permeability;
  - Extreme soil temperatures
  - Heat from pavement;
  - Soil pollution.
- **Plant-induced stressors:** These *stressors* shift the lens to how plants can create challenges for people in urban settings. Fallen leaves, pollen, or fruit may be viewed as potential problems that require constant clean-up; root systems can lift pavements or interfere with underground infrastructure, and dense vegetation may be associated with safety concerns. While these issues are not stresses for plants themselves, they strongly influence how plants are managed, constrained, or even excluded from urban spaces. Acknowledging these dynamics is crucial for developing more balanced strategies for urban greening.
  - Roots growth;
  - Falling leaves;
  - Pollens and other allergens;
  - Substances released by plants;
  - Possible fall of branches/trunks;
  - Shades caused by trees;
  - Branches to be climbed.

4.4 Discussion and Limitations

By combining visual data gathered during field observations with insights from the scientific literature in the natural sciences—particularly the work by Cavender & Donnelly [5]—, I developed a comprehensive list of *stressors* that should be considered as *inhibitors* when designing public urban spaces that involve plants. One key realisation that emerged (less evident during multispecies ethnography) is that while signs of stress often manifest in a plant's visible parts, particularly the leaves, the effects of stress extend to both above-ground and below-ground systems. Above ground, urban trees face limited access to sunlight due to the shadows cast by buildings, increased exposure to vehicle-related air pollution, and restricted space for crown expansion. At the same time, rising global temperatures and shifting climate patterns (such as heavy rainfall, extreme heat, and air pollutants) intensify pressures from pests, diseases, and invasive species [5]. Below ground, roots are subjected to stress caused by the artificial physical, chemical, and biological properties of urban soils, which differ significantly from those found in more *natural* environments [5]. One of the most critical issues is the intrusion into the Critical Root Zone (CRZ), the vital area surrounding the base of a tree where the roots responsible for its health and structural stability are located. Urban developments such as pavements, underground infrastructure, and building foundations are frequently constructed too close to this zone, severely compromising the plant's ability to absorb nutrients, water, and maintain overall resilience. This evidence (an issue that will be explored in detail in Chap. 5) reinforces the need for more plant-aware design practices that consider both the visible and invisible dimensions of plant life in urban contexts. This issue appears to arise mainly from biases such as the previously discussed *plant blindness*. Still, it is further compounded by humans' limited ability to perceive and engage with what happens below ground. Our nomadic way of life makes it challenging to fully comprehend how deeply plants are rooted in place and how crucial soil access is for meeting their diverse needs.

Since my knowledge of the basic information about the plants I encountered was limited, I found that key data (such as species, age, and CO_2 absorption capacity) is often readily available on dedicated online platforms. These are typically maintained by public administrations, indicating that the nonhuman heritage of urban areas is, to some extent, recognised and catalogued. However, my inquiry did not focus on assessing such data. Instead, it aimed to evaluate whether the agency of each plant encountered was being acknowledged and respected. While these platforms provide valuable knowledge, they fall short in capturing the lived, dynamic realities of plants. They do not reflect ongoing, day-to-day conditions or human-induced *stressors*, which can occur suddenly and have immediate impacts. A static image or data entry preserved for years cannot offer an accurate or current impression of a tree's health or well-being. Unlike buildings or artefacts, plants are living agents that continuously change in response to access to resources, environmental pressures, and human interactions. Thus, while helpful in comparing physical infrastructure, these digital mapping tools are insufficient for understanding the vitality and agency of plants.

4.5 Conclusions

Stressors are an almost constant feature of urban environments, particularly in public spaces such as streets and squares, where trees and other plants are continually shaped—and often constrained—by anthropocentric planning practices. The growing call to reimagine cities as multispecies contexts has begun to challenge the roles historically assigned to plants, prompting a reconsideration of how urban ecosystems are designed and managed. Although scientific literature and contemporary design practices increasingly explore innovative strategies for reshaping urban contexts, many of these approaches remain grounded in human-centred assumptions that overlook the agency, needs, and lived conditions of plants. At the same time, cities are becoming progressively more toxic and inhospitable due to the cumulative impact of environmental *stressors*, further complicating the survival and well-being of urban vegetation.

The insights gathered on plant-experienced *stressors*, their frequent neglect, and direct observations in the field raise questions about whether critical elements of a more-than-human perspective are being overlooked during urban planning processes, particularly at local scales. In neighbourhood planning, where decisions may not follow broader strategic frameworks, it is especially plausible that specific knowledge about plant agency and multispecies interactions is insufficiently considered. Addressing these intertwined challenges requires adopting a holistic and integrated approach that treats the city as an interconnected ecosystem from the outset, and that fully embraces its multispecies character as a foundation for more equitable and resilient urban futures.

References

1. BGCI (2021) The state of the world's trees. BGCI, Richmond
2. Nowak DJ, Greenfield EJ, Ash RM (2019) Annual biomass loss and potential value of urban tree waste in the United States. Urban For Urban Green 46:126469
3. Whyte WH (1980) The social life of small urban spaces. The Conservation Foundation
4. Czaja M, Kołton A, Muras P (2020) The complex issue of urban trees—stress factor accumulation and ecological service possibilities. Forests 11:932. https://doi.org/10.3390/f11090932
5. Cavender N, Donnelly G (2019) Intersecting urban forestry and botanical gardens to address big challenges for healthier trees, people, and cities. Plants People Planet 1:315–322
6. Dmuchowski W, Brągoszewska P, Gozdowski D et al (2022) Strategies of urban trees for mitigating salt stress: a case study of eight plant species. Trees 36:899–914. https://doi.org/10.1007/s00468-020-02044-0
7. Jim CY (2005) Outstanding remnants of nature in compact cities: patterns and preservation of heritage trees in Guangzhou city (China). Geoforum 36:371–385
8. Piana MR, Aronson MF, Pickett ST, Handel SN (2019) Plants in the city: understanding recruitment dynamics in urban landscapes. Front Ecol Environ 17:455–463. https://doi.org/10.1002/fee.2098
9. Mullaney J, Lucke T, Trueman SJ (2015) A review of benefits and challenges in growing street trees in paved urban environments. Landsc Urban Plan 134:157–166
10. Escobar A (2019) Habitability and design: radical interdependence and the re-earthing of cities. Geoforum 101:132–140
11. Caneva G, Bartoli F, Zappitelli I, Savo V (2020) Street trees in italian cities: story, biodiversity and integration within the urban environment. Rend Lincei Sci Fis Nat 31:411–417
12. Coccia E (2021) Métamorphoses. Wiley
13. Bodo T, Gimah BG, Seomoni KJ (2021) Deforestation and habitat loss: human causes, consequences and possible solutions. J Geogr Res 4:22–30
14. Vacchiano G (2019) La resilienza del bosco: storie di foreste che cambiano il pianeta. Mondadori, Milano
15. Antrop M (2004) Landscape change and the urbanization process in Europe. Landsc Urban Plan 67:9–26
16. Collinge SK (1996) Ecological consequences of habitat fragmentation: implications for landscape architecture and planning. Landsc Urban Plan 36:59–77
17. Clément G (2004) Manifesto of the third landscape. Editions Sujet/Objet
18. Coccia E (2018) La vita delle piante: metafisica della mescolanza. Società editrice il Mulino

19. Houston D (2019) Planning in the shadow of extinction: Carnaby's Black cockatoos and urban development in Perth, Australia. Contemp Soc Sci 16:43–56
20. Franklin A (2017) The more-than-human city. Sociol Rev 65:202–217
21. Berenguer J (2010) The effect of empathy in environmental moral reasoning. Environ Behav 42:110–134
22. Burayidi MA, Allen A, Twigg J, Wamsler C (2020) The Routledge handbook of urban resilience. Routledge
23. Owen D (2009) Green metropolis: why living smaller, living closer, and driving less are the keys to sustainability. Penguin
24. Haas T, Olsson K (2014) Transmutation and reinvention of public spaces through ideals of urban planning and design. Space Cult 17:59–68
25. Newman P, Jennings I (2008) Cities as sustainable ecosystems. Principles and practices. Island Press
26. Johnston P, Everard M, Santillo D, Robèrt K-H (2007) Reclaiming the definition of sustainability. Environ Sci Pollut Res Int 14:60–66
27. Wamsler C (2014) Cities, disaster risk and adaptation. Routledge
28. Rydin Y (2012) Governing for sustainable urban development. Routledge
29. Leitmann J (1999) Sustaining cities: environmental planning and management in urban design. McGraw-Hill Professional Publishing
30. Caves RW (2004) Encyclopedia of the city. Routledge
31. Lehmann S (2010) The principles of green urbanism: transforming the city for sustainability. Earthscan, London
32. Wolch JR, Byrne J, Newell JP (2014) Urban green space, public health, and environmental justice: the challenge of making cities 'just green enough'. Landsc Urban Plan 125:234–244
33. Hirons A, Thomas PA (2018) Applied tree biology. Wiley
34. Kuo M (2015) How might contact with nature promote human health? Promising mechanisms and a possible central pathway. Front Psychol 6:1093
35. Turner-Skoff JB, Cavender N (2019) The benefits of trees for livable and sustainable communities. Plants People Planet 1:323–335
36. Pearlmutter D, Calfapietra C, Samson R et al (2017) The urban forest: cultivating green infrastructure for people and the environment. Springer
37. Seamans GS (2013) Mainstreaming the environmental benefits of street trees. Urban For Urban Green 12:2–11
38. Tyrväinen L, Pauleit S, Seeland K, de Vries S (2005) Benefits and uses of urban forests and trees. In: Urban forests and trees. Springer, pp 81–114
39. Brink E, Aalders T, Ádám D et al (2016) Cascades of green: a review of ecosystem-based adaptation in urban areas. Glob Environ Chang 36:111–123
40. Cohen-Shacham E, Walters G, Janzen C, Maginnis S (2016) Nature-based solutions to address global societal challenges, vol 97. IUCN, Gland, pp 2016–2036
41. Ellis EC, Mehrabi Z (2019) Half Earth: promises, pitfalls, and prospects of dedicating Half of Earth's land to conservation. Curr Opin Environ Sustain 38:22–30
42. Frantzeskaki N (2019) Seven lessons for planning nature-based solutions in cities. Environ Sci Pol 93:101–111
43. Raymond CM, Frantzeskaki N, Kabisch N et al (2017) A framework for assessing and implementing the co-benefits of nature-based solutions in urban areas. Environ Sci Pol 77:15–24
44. Roggema R (2020) Nature-driven urbanism. In: Nature driven urbanism. Springer, pp 1–8
45. Xie L, Bulkeley H (2020) Nature-based solutions for urban biodiversity governance. Environ Sci Pol 110:77–87
46. Mollison B, Slay RM, Girard J-L, Girard J-L (1991) Introduction to permaculture. Tagari Publications, Tyalgum
47. Mollison B, Holmgren D (1978) Permaculture one: a perennial agricultural system for human settlements, 1990th edn. Tagari Publications

48. de la Bellacasa MP (2017) Matters of care: speculative ethics in more than human worlds. University of Minnesota Press
49. Clarke LW, Jenerette GD (2015) Biodiversity and direct ecosystem service regulation in the community gardens of Los Angeles, CA. Landsc Ecol 30:637–653
50. Pitt H (2017) Questioning care cultivated through connecting with more-than-human communities. Soc Cult Geogr 19:253–274. https://doi.org/10.1080/14649365.2016.1275753
51. Rice L (2018) Nonhumans in participatory design. CoDesign 14:238–257
52. Aziz H-A, Rasidi MH (2014) The role of green corridors for wildlife conservation in urban landscape: a literature review. In: IOP conference series: earth and environmental science. IOP Publishing, p 012093
53. Panagakos G (2016) Green corridors basics. In: Green transportation logistics. Springer, pp 81–121
54. Dunnett N, Kingsbury N (2008) Planting green roofs and living walls. Timber Press, Portland, OR
55. Oberndorfer E, Lundholm J, Bass B et al (2007) Green roofs as urban ecosystems: ecological structures, functions, and services. BioScience 57:823–833
56. Özyavuz M, Karakaya B, Ertin DG (2015) The effects of green roofs on urban ecosystems. In: Conference: GreenAge symposium. Mimar Sinan Fine Arts. University Faculty of Architecture, pp 15–17
57. Rowe DB (2011) Green roofs as a means of pollution abatement. Environ Pollut 159:2100–2110
58. Vijayaraghavan K (2016) Green roofs: a critical review on the role of components, benefits, limitations and trends. Renew Sustain Energy Rev 57:740–752
59. Liberalesso T, Cruz CO, Silva CM, Manso M (2020) Green infrastructure and public policies: an international review of green roofs and green walls incentives. Land Use Policy 96:104693
60. Manso M, Castro-Gomes J (2015) Green wall systems: a review of their characteristics. Renew Sustain Energy Rev 41:863–871
61. Solecki WD, Welch JM (1995) Urban parks: green spaces or green walls? Landsc Urban Plan 32:93–106
62. Heitlinger S, Foth M, Clarke R, et al (2018) Avoiding ecocidal smart cities: participatory design for more-than-human futures, pp 1–3
63. Kitchin R (2014) The real-time city? Big data and smart urbanism. GeoJournal 79:1–14
64. Tomitsch M, Fredericks J, Vo D et al (2021) Nonhuman personas: including nature in the participatory design of smart cities. Interact Des Architect 50:102–130
65. Zrajaeva I (2021) A tale of a wise city: A speculation on entanglements of non-humans and humans in an urban space. In: Brandt E, Markussen T, Berglund E, Julier G, Linde P (eds) Nordes 2021: Matters of Scale, 15-18 August, Kolding, Denmark. https://doi.org/10.21606/nordes.2021.14
66. Misiune I, Kazys J (2022) Accessibility to and fragmentation of urban green infrastructure: importance for adaptation to climate change. In: Misiune I, Depellegrin D, Egarter Vigl L (eds) Human-nature interactions: exploring nature's values across landscapes. Springer International Publishing, Cham, pp 235–246
67. Ershad Sarabi S, Han Q, L. Romme AG et al (2019) Key enablers of and barriers to the uptake and implementation of nature-based solutions in urban settings: a review. Resources 8:121
68. Abbott S (2021) Approaching nonhuman ontologies: trees, communication, and qualitative inquiry. Qual Inquiry 27:1059–1071
69. Diaz S, Settele J, Brondízio E, Ngo HT, Guèze M, Agard J, Arneth A, Balvanera P, Brauman K, Butchart S, Chan K, Garibaldi LA, Ichii K, Liu J, Subramanian SM, Midgley GF, Miloslavich P, Molnár Z, Pfaff A, Polasky S, Purvis A, Razzaque J, Reyers B, Shin JJ, Visseren-Hamakers IJ, Willis K, Zayas C (2019) The global assessment report on biodiversity and ecosystem services: Summary for policymakers. Intergovernmental Science-Policy Platform on Biodiversity and Ecosystem Services (IPBES), Bonn
70. Fry T (2017) Remaking cities: an introduction to urban metrofitting. Bloomsbury Publishing

71. Laybourn-Langton L, Rankin L, Baxter D (2019) This is a crisis: Facing up to the age of environmental breakdown. IPPR, London
72. Roudavski S (2021) Interspecies design. In: Cambridge companion to literature and the Anthropocene, pp 147–162
73. Vergani F, Tassinari V, Ferreri V (2022) Radical interdependence on a neighborhood scale. Raising awareness among children about human and more-than-human entanglements. In: Cumulus conference Detroit 2022 proceedings. Cumulus Association, pp 521–537
74. Camocini B, Fassi D, Meroni A (2017) In the neighbourhood and beyond. In: In the neighbourhood. Spatial design and urban activation. FrancoAngeli, pp 138–147
75. Sadeque S, Roy SK, Swapan MSH et al (2020) An integrated model of city and neighborhood identities: a tale of two cities. J Bus Res 117:780–790
76. Florida R (2002) The rise of the creative class. Basic Books, New York
77. Meroni A (2007) Creative communities. People inventing sustainable ways of living. Edizioni Poli.Design, Milan
78. Emilson A, Hillgren P-A, Seravalli A (2014) Designing in the neighborhood: beyond (and in the shadow of) creative communities. In: Ehn P, Nilsson EM, Topgaard R (eds) Making futures: Marginal notes on innovation, design, and democracy. MIT Press
79. Thorpe A, Manzini E (2018) Weaving people and places: art and design for resilient communities. She Ji J Des Econ Innov 4:1–10
80. DelSesto M (2020) People–plant interactions and the ecological self. Plants People Planet 2:201–211
81. Marder M (2013) Plant-thinking: a philosophy of vegetal life. Columbia University Press
82. Laist R (2013) Plants and literature: Essays in critical plant studies. Rodopi B. V., Amsterdam - New York
83. Hall C, Knuth M (2019) An update of the literature supporting the well-being benefits of plants: a review of the emotional and mental health benefits of plants. J Environ Hortic 37:30–38
84. Ostrom E (1990) Governing the commons: the evolution of institutions for collective action. Cambridge University Press, Cambridge; New York
85. Marttila S, Botero A, Saad-Sulonen J (2014) Towards commons design in participatory design. In: ACM, pp 9–12
86. Wall D (2005) Babylon and beyond: the economics of anti-capitalist, anti-globalist a. Pluto Press
87. Haraway D (1988) Situated knowledges: the science question in feminism and the privilege of partial perspective. Fem Stud 14:575–599
88. Sandler R (2018) Is considering the interests of plants absurd? In: Plant ethics. Routledge, pp 40–50
89. Višak T (2018) Utilitarian plant ethics. In: Plant ethics. Routledge, pp 30–39
90. Tsing AL (2015) The mushroom at the end of the world: on the possibility of life in capitalist ruins. Princeton University Press
91. Dowling R, Lloyd K, Suchet-Pearson S (2017) Qualitative methods II: 'more-than-human' methodologies and/in praxis. Prog Hum Geogr 41:823–831
92. Lien ME, Pálsson G (2021) Ethnography beyond the human: the 'other-than-human'in ethnographic work. Ethnos 86:1–20
93. Biggs HR, Rosén AP, Veselova E et al (2024) Multispecies ethnography in design research and practice. In: More-than-human design in practice. Routledge, pp 119–131
94. Kirksey SE, Helmreich S (2010) The emergence of multispecies ethnography. Cult Anthropol 25:545–576
95. Locke P, Munster U (2015) Multispecies ethnography. Oxford University Press, Oxford
96. Abbott S (2021) Tree knowing: ethnographic encounters, sensuous scholarship, relational ontologies, and environmental empathy. Royal Roads University
97. Reeves S, Kuper A, Hodges BD (2008) Qualitative research methodologies: ethnography. BMJ 337:a1020

98. Tsing AL, Bubandt N, Gan E, Swanson HA (2017) Arts of living on a damaged planet: ghosts and monsters of the anthropocene. University of Minnesota Press
99. Lynch J, Mannion G (2021) Place-responsive Pedagogies in the Anthropocene: attuning with the more-than-human. Environ Educ Res 27:864–878
100. Forlano L (2016) Decentering the human in the design of collaborative cities. Des Issues 32:42–54
101. Giaccardi E, Redström J, Nicenboim I (2025) The making (s) of more-than-human design: introduction to the special issue on more-than-human design and HCI. Hum-Comput Interact 40:1–16
102. Nicenboim I, Oogjes D, Biggs H, Nam S (2025) Decentering through design: bridging post-human theory with more-than-human design practices. Hum-Comput Interact 40:195–220
103. Gibson D (2018) Towards plant-centred methodologies in anthropology. Anthropol South Afr 41:92–103
104. Iwasa Y, Cohen D, Leon JA (1985) Tree height and crown shape, as results of competitive games. J Theor Biol 112:279–297
105. Kallarackal J, Roby T (2012) Responses of trees to elevated carbon dioxide and climate change. Biodivers Conserv 21:1327–1342
106. Fry T (2020) Defuturing. A new design philosophy. Bloomsbury, London

Chapter 5
Co-designing Public Spaces: Tools for Envisioning Multispecies Futures

Abstract Participatory design (PD) has traditionally focused on fostering democratic, inclusive, and community-driven design processes. However, in the context of the Anthropocene, PD is being reimagined to account for the entangled relationships between humans and nonhumans. This chapter extends the theoretical foundations laid in previous chapters by showing how the *Relational Atlas* (RA) can inform PD processes in urban public spaces. It describes an envisioning activity carried out in collaboration with a group of local stakeholders, centred on a small urban public space where plants come across multiple *stressors*. The goal of this activity was to raise human awareness and stimulate the development of multispecies collectives. Drawing on speculative design and design fiction, participants were asked to represent plant species, challenging anthropocentric assumptions and incorporating plant perspectives into the design process.

5.1 Introduction

The previous chapters have provided a theoretical foundation for exploring more--than-human and multispecies perspectives. The *Relational Atlas* (RA) served as a means of bringing together knowledge and concepts from multiple disciplines into a tool that not only seeks to make sense of interspecies relations as a cornerstone of more-than-human design but also aspires to establish a framework for designing multispecies experiences. Nevertheless, as emphasised in the previous chapter, the RA remains a tool primarily accessible and helpful to humans.

Within the field of participatory design (PD), such theoretical frameworks are valuable in preparing human communities by offering scientific insights into non-human agency. Although other species cannot directly use them, they lay the groundwork for imagining more open-ended actions, tools, and experiences.

This chapter builds on the theoretical framework previously discussed and illustrates how the collected knowledge proved essential in rethinking the design of a small urban public space. The PD activity presented here is particularly concerned with raising human awareness around envisioning multispecies communities, with

F. Vergani, *Reframing Urban Ecologies*, SpringerBriefs in Applied Sciences and Technology, https://doi.org/10.1007/978-3-032-18924-0_5

special attention to plants that are currently suffering from a series of *stressors* (the inhibitors framed in the *Atlas*). While not fully aligned with more-than-human methodologies—leaning more toward a *design for* rather than a *design with* approach—the case study critiques the typically anthropocentric orientation of design. Through speculative design and design fiction, it encourages the human community to acknowledge nonhuman agency by assuming the role of nonhuman representatives, thereby embracing plants in the design process.

5.2 Participatory Design and Nonhumans

Designing *for* and *with* multispecies communities in cities means engaging with complex, situated contexts [1] shaped by a plurality of actors [2] and spaces governed by diverse political and administrative bodies. Within this framework, participatory design (PD) serves as a design approach that co-creates processes and outcomes from a community-centred, wide-ranging perspective [3, 4], thereby expanding the democratic arena to encompass the diverse perspectives of all participants [5]. PD is a reflective practice grounded in the principles of inclusion, ethics, and democracy [2, 6–11], closely linked to themes such as social, environmental, and "multispecies justice" [12, 13], as well as "activism" and "agonism" [2, 9, 14–20].

Hence, Design, and particularly Design for Social Innovation, "has become a formidable political and material force" [21], playing a crucial role in generating new entities and relations among more-than-human collectives. In response to the challenges of the Anthropocene and the issues previously discussed, PD must reframe its political role by acknowledging the thick entanglements [22] within more-than-human communities, moving beyond the exclusively human sphere and expanding the political agency traditionally envisioned for PD [8, 9] toward a (cosmo)political perspective [23]. This means reimagining PD approaches of *infrastructuring*, *commoning*, and *institutioning* [24–26] through an ontological lens [6, 27]. In recent years, however, scholars have raised concerns about the potential drawbacks of PD and other human-centred approaches, which prioritise the well-being and needs of humans above all else [28, 29]. As Forlano [28] argues, we must "decenter the human to consider the agency of nonhumans in the context of the Anthropocene." In this light, PD requires a shift from a human-centred paradigm toward a more-than-human lens to embrace the perspectives of the nonhuman realm along with their corresponding ethical, legal, and methodological implications [30].

Although the idea of extending PD to include nonhuman agents was theorised over a decade ago [31, 32], research and practice in this direction remain relatively uncommon [6, 33–39]. Within this context, Roudavski [40] has articulated participation with nonhumans through the concept of interspecies design, defined as "a form of design that seeks to involve and benefit both human and nonhuman life-forms." Emerging primarily from ethnographic research methods aimed at identifying collaborations between different agents, interspecies design emphasises the need to embrace the distinct qualities of each species to foster an "interspecies culture" [40–42] that supports all stakeholders involved [7, 43, 44]. To date, these multispecies approaches of PD have been applied more frequently in animal–human collaborations [45–50] than in plant–human cooperation [33, 37, 44].

However, implementing these multispecies forms of PD in practice proves highly complex for two main reasons. First, while research on plants has identified principles for potential interspecies participation, these have yet to be fully realised [7, 51–53]. The lack of adequate technological systems and devices to facilitate interspecies communication, coupled with the inherently ambiguous nature of communication across species, forces the scientific community to seek alternative ways of integrating plant knowledge into PD processes, also relying on indigenous practices.

Within this context, speculative design [54] or critical fabulation [55, 56] may offer practical approaches for addressing the lack of nonhuman participation. They can temporarily bridge these gaps by inviting experts, "representatives" [57], "spokespersons" [58] or using devices that act as "mediators" [38, 59, 60], to convey the *voice* of nonhumans. Second, as already mentioned throughout the volume, a significant perceptual bias persists among human stakeholders in recognising plants as equally important agents [38, 40].

5.2.1 Nonhumans as Citizens and Stakeholders in Participatory Processes

Designing from a more-than-human perspective requires engaging with diverse communities in which plants are recognised as agents, even as "persons" [61]. As Gibson and Baylee [62], articulating on Gagliano et al. [63] and Hall [61], highlight, "plants do not resemble people, but new ecological work offers a way of approaching plant life in personal terms: through experience, relationality, perception, language, and intelligence." Indeed, plants are inhabitants of cities, and there is an increasing call to recognise plants and trees as *citizens* within urban discourse [8, 64]. According to the *Britannica Encyclopedia*, citizens are individuals who possess specific rights, duties, and responsibilities, as well as political and economic ties to the country in which they reside. Nyers [65] further defines citizens as "individuals with an audible and bodily presence," arguing that citizenship should be understood not as a fixed status but as an ongoing process that ought to remain open to those who are often excluded, such as migrants and refugees. In this context, Häkli's [66] concept of "positional citizenship" offers a lens through which to rethink the very notion of citizenship, bridging the ontological divide between humanist and posthumanist thought as a step toward reimagining civic life in a multispecies world. From this perspective, nonhumans can also hold a *position* in society, as they too are subject to forms of exclusion and discrimination, much like other marginalised human groups. This idea resonates with Hannah Pitt's [44] notion of *ecological citizenship*, which frames citizenship as an ethical practice in which individuals and communities actively engage in the care, protection, and stewardship of the *natural* environment. Pitt argues that traditional notions of citizenship, typically centred on political rights and responsibilities, must be broadened to encompass an ecological dimension, thereby including other nonhumans. She further emphasises that ecological citizenship requires not only individual choices but also structural transformations, such as policies and institutions that enable and sustain environmentally responsible practices. Yet, as previously noted, plants and other nonhumans still await recognition as political agents, as they remain without formal acknowledgement or rights. Only a handful of municipalities worldwide have begun to move beyond the strictly political and economic dimensions of citizenship, expanding their policies to embrace nonhumans that play a vital role in the ecosystem.

The concept of participation in design processes enables the translation of citizenship, still too ambiguously defined, into a more fluid notion of *stakeholder*, or *shareholders*, to use a new concept coming from economics. Among others, Gibson and Baylee [62], Roudavski [40], Thomas et al. [67], and Tomitsch et al. [60] discuss this shift, while Veselova and Gaziulusoy [38] argue that a nonhuman can qualify as a stakeholder if it is an agent "without formal design training who can inform and is involved in or affected by the design process and its outcomes." Considering nonhumans as stakeholders in PD practices for public urban spaces also requires engaging with the convergence of top-down and bottom-up dynamics. Tomitsch et al. [60] describe this as a "middle-out engagement," which brings together representatives from top-down and bottom-up organisations in community engagement to form a coalition capable of speaking on behalf of nonhumans. Within this framework, PD tools can help address the complexities of designing with nonhuman communities. For instance, personas, a well-known service design tool, can serve as "an effective way to make sense of and synthesise research data, to communicate user needs within the design team, and to keep the perspective of users and other important stakeholders at the forefront throughout the design process" [60].

5.2.2 *Middle-Out Engagement in City Neighbourhoods*

As already introduced in Chap. 4, neighbourhoods are increasingly becoming laboratories for experimenting with new approaches to city design and addressing the social, environmental, and cultural challenges they face [68]. Due to their relatively small scale, they are well-positioned to tackle these complexities from a community-centred perspective, with citizens serving as the primary actors who experience such issues firsthand on a daily basis [43]. Communities are engaging more actively with their local contexts, developing ideas and practices that both innovate and safeguard common goods, while fostering inclusive and democratic processes of environmental, economic, and social regeneration [69, 70]. Within neighbourhoods, the most familiar example of a common good is the public space, broadly defined as open, accessible areas such as streets, parks, and squares, typically managed by governmental or public entities. These spaces link private lots and buildings, providing places where agents gather, socialise, and engage in a wide variety of activities [71–73].

Although municipal administrations oversee the physical management of public spaces, citizens can play an active role in their upkeep. As Deshpande [74] notes, citizens have a "moral responsibility" to care for these spaces and to foster civic involvement and community engagement inclusively. The actions of individuals (both human and nonhuman) inevitably influence the condition of spaces that are, by definition, shared and experienced by all. However, citizens often find themselves facing challenges for which they are not fully equipped to respond effectively.

Approaching transformative pathways in neighbourhoods requires the previously mentioned "middle-out engagement", spanning from grassroots initiatives driven by local groups to top-down actions supported by institutional bodies [68, 75]. In the last decades, local administrations have recognised the value of neighbourhood communities, aided by new administrative tools, such as *participatory budgeting* or *collaborative pacts*, that strengthen citizens' roles in developing and co-designing solutions and experiences for their communities. These tools foster temporary design approaches in which "creative" [4] or "project-based" [75] communities are directly engaged in revitalising the public spaces of their neighbourhoods. However, the complexity of this process cannot be addressed from a single perspective; it requires building a network of diverse agents from multiple domains. Citizens, local businesses, associations, informal groups, policymakers, institutions, municipal administrations, and experts [75] should come together to form "coalitions" [60] capable of tackling these challenges more effectively. Within this framework, designers can play a pivotal role, acting both as process facilitators and content creators [76]. More broadly, design can contribute to shaping policies and administrative regulations by drawing attention to specific scenarios and sparking new perspectives on complex issues [77].

However, when it comes to the nonhuman members of a community, we still lack effective methods to actively involve them in community-making processes, as Tomitsch et al. [60] suggest, the middle-out engagement approach can also be applied to integrate nonhuman agency by engaging experts, ecologists, conservationists, and similar professionals who can bring their perspectives into the discussion. These actors may assume the role of "representatives" [57], providing data and insights about specific nonhumans in order to strengthen the coalition. However, although these methods allow plants, animals, and other species to be acknowledged as stakeholders, it remains challenging to speak of genuine and fully realised participation among multispecies agents. Within this framework, design can help strengthen the human side of the community by envisioning new scenarios and possible transformative pathways that address the needs of nonhumans, fostering mutual benefits for all agents involved [7, 44, 53].

5.2.3 Social Innovation and Participatory Design

As previously discussed, within the many branches of design, Design for Social Innovation holds particular potential for fostering social cohesion, empowering citizens to address the challenges of urban living, and envisioning projects of all kinds [78]. It can help create inclusive, accessible spaces that encourage community engagement and inspire people to take an active role in shaping their surroundings. Rooted in PD as its primary approach, Design for Social Innovation also draws on PD-specific tools and artefacts to co-create situated but plural outcomes. According to Fassi and Sedini [77], citing Manzini [2], social innovation solutions are "all social changes towards sustainability [that] can reduce environmental impact, regenerate common goods, and strengthen the social fabric." These solutions are typically tailor-made by designers who work within, and actively participate in, specific communities of people, the "situated stakeholders" [68], who safeguard the well-being of their context by caring for shared resources.

PD plays a pivotal role in expanding the democratic arena by embracing the diverse perspectives of participants [5] while navigating the divergences and complexities inherent in situated multispecies communities. Communities are, in essence, spaces where multiple *voices* converge, forming a "pluralistic" context [79] in which "agonism" [16, 80, 81] comes into play. Through the process of sharing and debating these varied viewpoints, PD can help uncover shared "matters of care" [82]: issues that, at their core, concern us all [83]. Such matters of care are also embedded in the intricate entanglements between human and nonhumans, entanglements that have been progressively eroded over centuries by anthropocentric decisions. From the knowledge we have on trees (as discussed in Chap. 4), such choices manifest as *stressors* that undermine their well-being.

PD must move beyond the human-centred realm and adopt an ontological perspective [27] that rethinks the relationships between diverse entities and categories of being, to acknowledge the "radical interdependence" [21] that binds all agents, and to imagine scenarios for multispecies coexistence within the situated contexts [1] in which we operate.

5.3 Participatory Design into Play: Granting the Involvement of Plants in Redesigning a Public Space

The theoretical background presented explores themes such as more-than-human citizenship and participation in urban space design, emphasising the need to focus on situated urban contexts when discussing relationships. In this light, while the *Relational Atlas* addressed the agency of plants from both biological and relational perspectives, the work presented in this chapter investigates the *social status* of plants as perceived by human citizens. This is where PD comes into play, testing its capacity to envision innovative scenarios and shifting the focus from a human- or community-centred approach toward a more-than-human, multispecies perspective. Facilitating dialogue among the diverse and distinctive *voices* within a situated community can help foster a shared neighbourhood culture and inspire future actions that acknowledge and integrate common matters of care. In doing so, the chapter lays the groundwork for a more-than-human community that embraces its multispecies dimension.

To pursue these broader objectives, I set out to test the insights derived from Critical Plant Studies—outlined in the previous chapters—through a co-design session within a specific situated community I had engaged with a few years earlier, where themes such as *multispecies justice* [12, 13] and the importance of *green spaces* were pivotal. Rooted in the theoretical framework of the RA, this activity seeks to address the *stressors* identified during an on-field multispecies ethnography in Nolo, Milan. It was implemented through a *provotype* [84, 85], a purposefully designed tool intended to mediate discussion and envision scenarios of coexistence. This approach embraced plants as equal stakeholders in the dialogue, with their perspectives conveyed through representatives [57] who acted as their *voices*.

This activity draws on design fiction [86, 87] and speculative design [54] to explore future possibilities and challenge current perspectives. Design fiction employs speculative thinking and storytelling to imagine and *provotype* potential futures, encouraging reflection on how design decisions influence how we live, work, and relate [88, 89]. Widely used in foresight, scenario planning, and co-creation, it examines the impacts of emerging technologies, policies, and trends while questioning prevailing assumptions [90]. Speculative design similarly imagines alternative futures to provoke critical reflection [54] and inspire transformative thinking [91, 92]. The purpose of this work was to apply these methods to consider plants as stakeholders, aiming to reframe human–plant relationships despite the current limitations in knowledge and tools.

5.3.1 The Context: Piazzetta Transiti

The co-design session took place in Nolo, a semi-peripheral neighbourhood of Milan undergoing significant transformation. Since 2015, Nolo has undergone a "beautification" process driven by an active and engaged community that has initiated a series of initiatives fostering social cohesion and innovation [68, 93]. This grassroots movement attracted the attention of the Polimi DESIS Lab research group at the Department of Design (Politecnico di Milano), of which I am a member. Beginning in 2016, the research group initiated a series of teaching and research activities in the area, culminating four years later in the establishment of Off Campus Nolo, a living lab housed in the neighbourhood's historic covered municipal market. Off Campus Nolo is part of a larger initiative developed by the Politecnico di Milano to make its presence in the city more tangible, allowing researchers and academics to be more responsible, sensitive to social challenges, and connected to the situated community and territory. The living lab supports not just research initiatives and

Fig. 5.1 *Piazzetta Transiti* (Transiti Square) in Nolo, Milan (Italy)

community-building practices, but also neighbourhood-organised events, meetings, and volunteer activities.

Within this framework, one of the key spaces where we collaborated with the Nolo community was *Piazzetta Transiti* (Transiti Square), a small urban park widely perceived as degraded, mainly due to social tensions involving marginalised immigrant groups [83] (Fig. 5.1). Over the past decade, repeated frictions among local stakeholders prompted calls for municipal intervention, which led to the removal of public furniture (primarily benches) and the installation of a tall metal fence. This measure restricted access not only to the so-called "night community" deemed responsible for the degradation, but also to long-standing users, particularly elderly residents. The installation process also involved clearing several bushes along the perimeter, further altering the park's ecological characteristics. While initially welcomed by some as a solution, the intervention effectively stripped the square of its role as a shared public space, reducing it to a mere transit corridor. The night community's relocation to neighbouring sidewalks and parking lots increased social tensions, attracted non-residents looking for cheap alcohol from nearby shops, and eventually reinforced impressions of neglect and insecurity. To reactivate *Piazzetta Transiti*, we applied to a municipal public call for revitalising spaces through a *collaborative pact*, an institutional tool supported by the Milan Municipality. We began by building a coalition of situated stakeholders—proactive citizens, informal groups, and associations previously engaged through Off Campus Nolo—committed to rethinking the space.

Through field meetings, we explored diverse perspectives and co-envisioned possible futures for the square. Parallel to these negotiations, the coalition organised events to test activities and encourage shared use, restoring its role as a common good. One summer event for children began with a community-led cleanup, removing bottles, caps, and other litter left by night regulars. Continued meetings strengthened commitment, and in February 2022, the coalition formalised its goals (making the space greener, safer, more inclusive, and vibrant) by signing the *collaborative pact*. The agreement grants 3 years of shared stewardship, enabling direct action without prolonged bureaucratic delays, thanks to the partnership with municipal administrations, who were also signatories.

5.3.2 Nonhumans in Transiti

The ongoing work at Off Campus Nolo—and particularly one of its flagship research projects, the *Situated Vocabulary* [83, 94]—centres on themes of community engagement, activism, and democracy. These issues can be addressed through the lens of PD, which invites diverse actors to envision new paths for their collective interests. More broadly, the project also engages with concepts such as "multispecies justice" [12, 13] and the need to recognise the agency of plants as a step towards designing more ecosystemic experiences that embrace all agents within a given context. As in the case of *Transiti*, specific neighbourhood spaces are rich in pluralities and diversities, and can be described as "friches" [95] or urban "wastelands" that also serve as refuges for marginalised agents, both human and nonhuman.

Primarily a garden, *Piazzetta Transiti* hosts several trees whose overall condition, despite occasional damage from the surrounding sidewalk, appears stable. Its vegetal heritage includes two magnolia grandiflora, five Norway maples, three tilias, two Judas trees, two red oaks, and two hornbeams (Fig. 5.2).

However, over the years, several newly planted trees have died for reasons that remain unspecified. According to the *stressors* identified in the previous chapter, trees in this area may be affected by artificial light, pollution from vehicles and infrastructure, high air temperatures, restricted root space, poor soil composition and structure, and climate change.

This information, gathered through ethnography, proved essential for assessing the health of the plants and, in turn, for setting up the co-design session. Based on the premise that plants should be recognised as citizens in their own right, their agency must not be disregarded. The primary aim of the activity was therefore to address their suffering by developing a set of tools to translate their needs and mitigate the inhibitors, paving the way for a renewed redesign of the area.

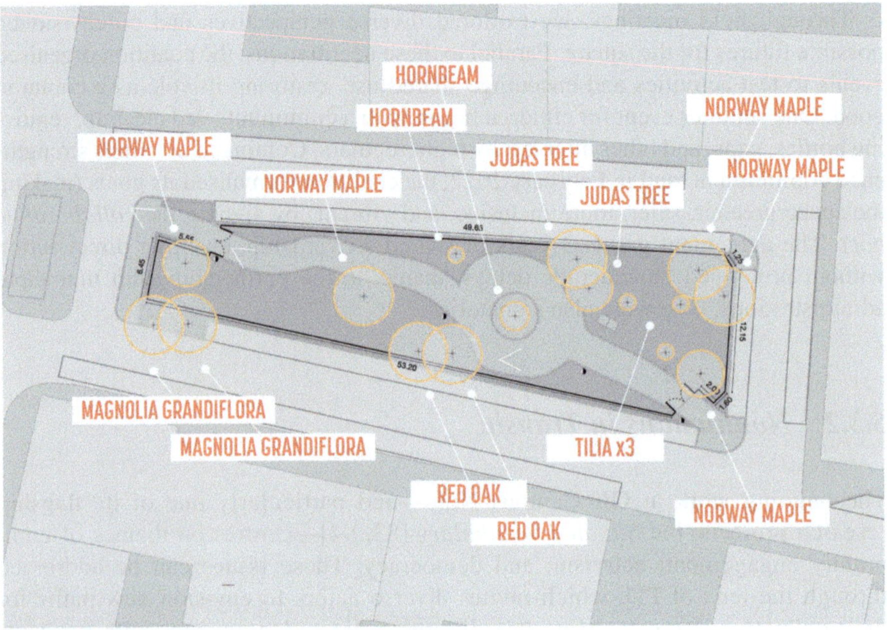

Fig. 5.2 All the trees living in *Piazzetta Transiti*

5.3.3 The Co-design Session

To address the plant inhibitors in *Transiti*, the co-design session combined elements of design fiction, speculative design, and spatial design. The activity was supported by a dedicated toolkit [86] containing probes and a purpose-built *provotype* [84, 85] designed to mediate discussion, envision scenarios of coexistence, and treat plants as equal stakeholders in the debate, even if represented through human proxies [57] who conveyed their *voices*. Indeed, as Mogensen [85] notes, a provotype "can serve as a bridge between analysis and design [as] it uses the results of analysis [to construct] first guesses in a prototyping process by providing ideas as to what should be changed and what should remain." The provotype was conceived using data gathered through the RA, combined with foundational knowledge in politics and law. Inspired by Latour's concept of the "Parliament of Things" [96] (a metaphor for a society in which both humans and nonhumans are represented in political processes), I developed a fictional narrative set in the year 2048, envisioning a future where plants have finally been granted political and legal rights. While the whole backstory was not explicitly revealed to participants, it was subtly embedded and could be inferred from the tools provided during the PD activity. These purpose-designed tools for the session included:

- ***Action Cards*** (Objective: Co-envision the scenarios for the Pact). The cards were divided into four thematic groups, serving as prompts for co-envisioning multiple scenarios to be developed within the future timeframe of the pact. The coalition was tasked with selecting actions for each scenario from four categories: "Actions of Care," "Actions of Regeneration," "Actions of Integration," and "Actions of Removal." While the cards were designed to remain generic and adaptable to other contexts, some were explicitly tailored to *Piazzetta Transiti*, drawing on data collected during previous months of meetings as well as findings from earlier chapters. In addition to strictly human-oriented actions, several cards addressed plant-specific needs (such as watering routines or installing curbs to protect root zones) alongside more contentious proposals that had sparked debate in earlier stakeholder discussions, like the installation of surveillance cameras to monitor the area. These cards were intentionally included to sustain an agonistic negotiation among diverse participants, aiming to foster democratic debate through the disarticulation and re-articulation of polarising viewpoints toward a shared objective. Ultimately, the creation of these future scenarios was designed to catalyse strategic conversations among stakeholders, supporting participatory and informed decision-making processes [92].

- ***Axonometry of Transiti Square*** (Objective: Co-envision the scenarios for the Pact). Three axonometric drawings illustrating the current state of *Piazzetta Transiti* were provided as a foundation for sketching future scenarios based on the selected cards. These scenarios were developed in response to three key challenges distilled from insights gathered during previous meetings:

  - Preventing inappropriate or disruptive use of the space during nighttime.
  - Enhancing the livability of the area, beginning with basic cleaning measures and extending to improved safety.
  - Levelling the *Piazzetta* with its surroundings to address issues of accessibility and overall spatial usability.

  The challenges were structured across varying time frames, beginning with less structured, socially driven commitments and progressing toward the complete redesign of the square. This phased approach was intended to enable the coalition to envision the most desirable of the possible futures, [97], as identified by the community in earlier meetings.

  Generic folder containing (Fig. 5.3)**:**

- ***Plant's Bill of Rights*** (Objective: Convey the backstory to the coalition). Inspired by the fictional *Plant's Bill of Rights* proposed by botanist Mancuso in *The Nation of Plants* [98] and journalist Viola in *Flower Power* [99], I created a document set in 2048 that mirrors the structure of the first Italian Constitution. Its purpose was to familiarise participants with the (fictional) rights granted to plants by the Italian government. While not central to the activity, this document served as a prop to enrich the backstory and provide context for the subsequent material.
- ***Legal forms*** (Objective: Convey the backstory to the coalition/Provide the challenge). Drawing inspiration from real legal documents used by prosecutors to initiate legal action, I created five fictional complaints written from the plants' perspective. Through the lens of design fiction, I imagined that these vegetal beings, following the declaration of the *Plant's Bill of Rights* and the develop-

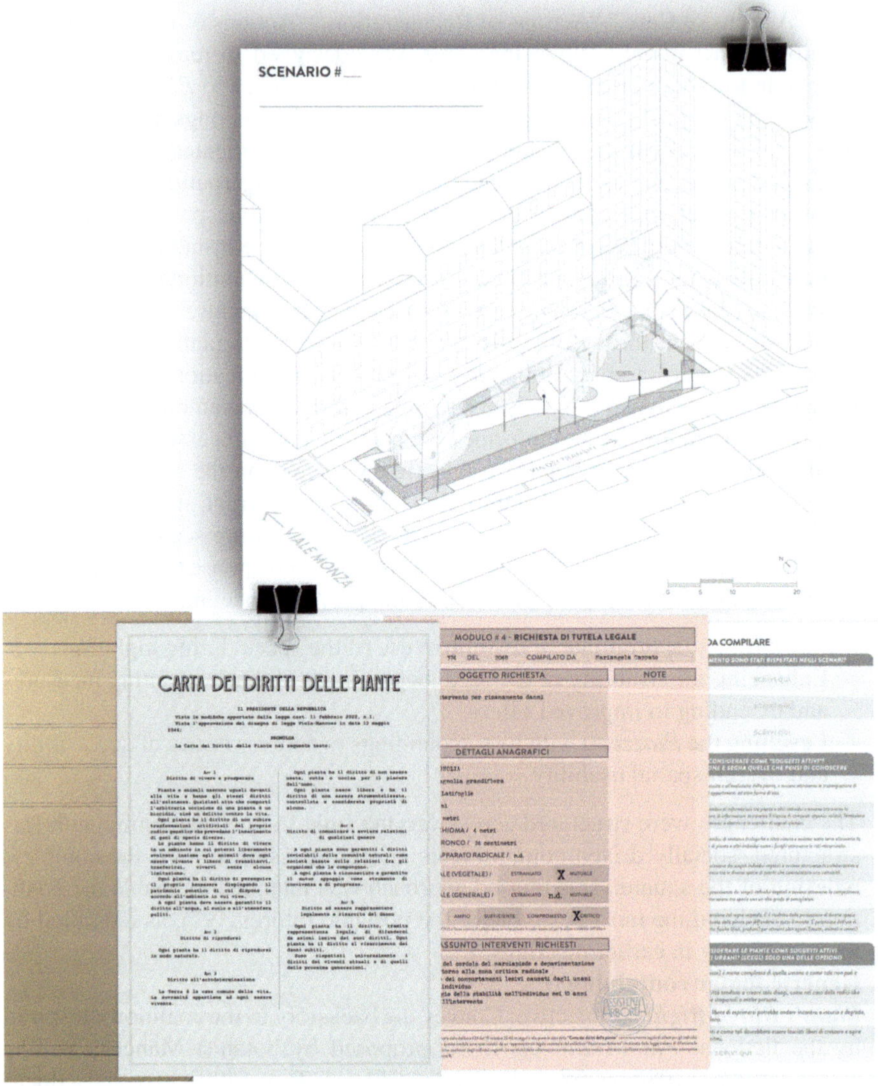

Fig. 5.3 The documents given to the participants. From the left: (1) *Piazzetta Transiti's axonometry*; (2) *Plant's Bill of Rights*; (3) *Legal Forms*; (4) *Feedback forms*

ment of an interspecies "communication" system (implied through brief notes at the bottom of the document but not described in detail), had decided to sue humans for harmful actions endured over the years. In this narrative, the five plants sought compensation for these injustices by engaging *Assistenza Arborea* ("Tree Care"), a fictional collective dedicated to defending their rights, which also drafted the complaint forms provided to human participants. Each fictional case was tailored to the actual trees inhabiting *Piazzetta Transiti* and reflected their specific needs:

- ***Magnolia grandiflora*:** requesting the removal of the curb and the depaving of the area around its critical root zone, an end to harmful human behaviours, and stability monitoring for the 10 years following the intervention.
- ***Tilia*:** soon to be planted on-site, expressing concern over insufficient care and asking for a reevaluation of the current project.
- ***Judas Tree*:** advocating for the creation of a *wildlife island* on-site to encourage pollinator proliferation.
- ***Norway maple*:** calling for improved tree care during hot seasons and the cessation of unnecessary plant felling.
- ***Red Oak*:** demanding legal action against indiscriminate branch pruning.

This specific *provotype* draws significant inspiration from the concept of *personas*, archetypes developed from data gathered through close observation of potential users and stakeholders [100, 101]. To structure the information presented in the form, I adapted a standard template that typically includes elements such as *backstory, motivations, frustrations, ideal experience, goals, aspirations,* and *feelings.* However, in this case, the template was reworked, and its structure was made less explicit through the use of the legal action metaphor, which reframed the content more imaginatively and indirectly.

The *legal forms* include the following information:

Side 1 /

- ***Form information*** (module number, year, name of the legal representative completing the form): establishes the fictional timeline, offering participants an initial clue about the future-oriented scenario in which plants have been granted legal rights.
- ***Request***: specifies the intervention sought by the plant.
- ***Personal details*** (Name, Species, Family, Age, Crown Diameter, Trunk Diameter, Root Diameter): provides both taxonomic and dimensional data, reinforcing the plant's individuality and value beyond species classification alone.
- ***Relational status(Vegetal)***: describes the plant's capacity to communicate with other plants to inform participants about the dynamics of root systems in urban public spaces.
- ***RelationalsStatus(General)***: outlines the plant's ability to communicate with agents beyond its own species, further contextualising the complexities of its interactions in shared environments.
- ***Critical Root Zone status***: highlights the condition of the plant's root system, emphasising potential harm caused by restricted space in urban settings.
- ***Request summary***: condenses the core intervention sought.
- ***Notes***: left blank for any additional remarks or clarifications.
- ***Footnotes***: provides the only element explicitly hinting at the fictional context. Written in formal, bureaucratic language, they provide details about the role of *Assistenza Arborea* in transcribing plant requests via a specialised communication system.

Side 2 /

- ***Request***: details the specific intervention sought by the plant, incorporating the fictional narratives created to support its legal claim.
- ***Location***: indicates the exact position of the plant within *Piazzetta Transiti.*
- ***Notes***: provides additional context on the request, including witness statements that validate and support the plant's case.
- **Feedback Form** (Objective: Collecting feedback from the participants): a form designed to capture participants' reflections on the co-design session. The form was structured around three key questions:
  - Were the trees' requests respected in the scenarios?
  - Can plants be considered "active subjects"? Read the following strategies and mark the ones you are familiar with.
  - What is your opinion on considering plants as active subjects in the design of urban spaces? (Select only one option.)

5.4 Wrapping Up the Results

The co-design session, held at Off Campus Nolo, involved five members of the coalition and lasted approximately 2 hours (Fig. 5.4). After a little ice-breaking, I introduced the three key challenges, reinforcing the concept of the different envisioned time spans. I then placed the three blank scenarios (axonometric drawings of *Piazzetta Transiti* printed on paper and mounted on small wooden panels) on the table alongside the deck of "Actions" cards. Following a brief introduction, I divided the deck and distributed groups of cards to the participants, asking them to assign each card to the scenario they felt best matched. Before starting the design phase, I handed out folders containing the *provotypes* (*The Plant's Bill of Rights* and the *Legal Form*) and explained the additional challenge: acting as a representative for a specific tree, ensuring that its requests were reflected in all three scenarios. Participants took time to read the forms, asking questions about particular terms that had been intentionally left only partially explained (such as *Critical Root Zone* or mycorrhizal networks) so I could provide clarifications when needed. As they began the co-design session, I reminded them to keep their "prosecutors'" (trees') requests at the forefront, avoiding solutions that might cause negative short- or long-term impacts.

At the end of the session, the three scenarios were only partially completed. Certain cards sparked extended discussions (for example, whether to install rubbish bins or convert the adjacent street into a pedestrian area) while other actions had to be added using pre-printed blank cards. In several cases, cards were reconsidered due to conflicts with the plants' requests outlined in the *provotypes*. Notably, during the co-design process, the following design decisions were revised:

Fig. 5.4 Pictures taken during the co-design activity with the coalition of Nolo *situated stakeholders*

- **Scenario 1**: the proposal to add extra night lighting as a deterrent for nocturnal visitors was abandoned, as artificial light can disrupt the biological rhythms of trees and cause stress.
- **Scenario 2**: the location chosen for a new fountain was moved to prevent potential damage to the roots of a Judas Tree, following a recent incident where soil removal for analysis had harmed them.
- **Scenario 2**: the idea of installing a small permanent stage for concerts and events was reconsidered, as it could obstruct rainwater infiltration into the soil. The coalition instead opted for temporary structures or porous materials.
- **Scenario 3**: the layout for a new walkway, intended to replace the current one, was modified to avoid disrupting communication through the mycorrhizal networks.

During the session, we also addressed the recurring issue of replacing dead plants in *Piazzetta Transiti*. While preparing the narrative for the Maple (Form 4), I used Google Earth's historical imagery to compare photographs from previous years, confirming that two trees of the same species had been replanted in the same spot within 2–3 years of each other. Although the cause of death remains unknown, the evidence of such a short lifespan underscores the importance of prioritising "Actions of Care" such as *watering* and *citizen science* to monitor the health of trees in the square regularly.

5.5 Discussion and Limitations

By incorporating the data previously gathered for the RA, I was able to introduce knowledge that was, at least in part, unfamiliar to the Nolo community. Participants responded attentively to the framing of plants as stakeholders, benefiting from essential information delivered in an alternative, less didactic manner. Members of the coalition appeared receptive to the topic and willingly embraced the challenge of considering plants' needs not only in the immediate design process but also in future maintenance.

The results indicate that PD and its tools can play a meaningful role in recognising plants as stakeholders in the design of public urban spaces. This potential stems largely from PD's ability—particularly when combined with approaches such as design fiction and speculative design—to engage with this complex topic in ways that resonate with communities. Participants showed strong interest in the activities, mainly because the process reframed the perspective from human-centred to more-than-human design. One of the most challenging aspects to convey, however, was how to address the specific needs and issues faced by plants in urban environments. In earlier meetings, before the signing of the *collaborative pact*, plants were generally regarded as static urban furnishings, comparable to benches or streetlights, valued for their ability to enhance the visual appeal of public spaces and counteract degradation. Their actual needs, along with the environmental *stressors* they endure daily, received minimal attention. This was partly because, at the outset, the primary focus was on forming the coalition and identifying the negative aspects of the area that required intervention.

The *provotype*, however, proved effective in exposing the harmful dynamics of human-centred design within public urban contexts. It allowed me to support the community of situated stakeholders in avoiding (or eliminating) design proposals that could, over time, harm the local plant community in *Piazzetta Transiti*. Particular emphasis was placed on the emerging idea of granting plants legal rights, a topic that genuinely captured the interest of participants. They especially appreciated how the fictional narratives illuminated issues that are often overlooked in urban planning. Overall, thanks to a collective discussion at the end of the activity, as well as written feedback gathered through the feedback form, the PD activity was well received by the coalition, demonstrating its value as both a practical and reflective tool for advancing multispecies-inclusive design.

The activity carried out through the provotype offers both a situated contribution to the *Transiti* project and a broader significance for PD practices. In the context of *Transiti*, the process suggests that a "middle-out engagement" approach can be an efficient strategy. Within this framework, design can play a pivotal role in translating insights from Critical Plant Studies into urban contexts, activating processes and practices that envision, and work toward, future scenarios of human–nonhuman coexistence. Moreover, design can support the work of policymakers and other administrative bodies, as research-driven projects can draw attention to specific scenarios that inspire new perspectives on pressing challenges. More broadly, applying PD within heterogeneous communities can be strengthened by the imaginative potential of speculative design and design fiction, which expand the space of possibilities and open the way for inclusive, multispecies futures. However, in bottom-up design sessions such as this one, situated communities should be equipped with targeted tools that prompt them to consider the needs of nonhumans throughout. Ideally, this awareness would be reinforced through legal recognition; however, such measures are unlikely to be adopted by governments and policymakers in the near future.

Several limitations emerged from the co-design activity. First, the *provotypes* were informed exclusively by secondary sources rather than primary empirical data, which may have reduced their scientific robustness. This limitation reflects a broader challenge in research involving nonhuman stakeholders, where direct forms of inquiry are inherently constrained. Second, the composition of the coalition introduced another limitation. The participants involved were notably attentive to the needs of plants: an attitude that may not reflect the perspectives of a broader population. To obtain a more representative understanding, additional participatory design activities should be conducted with more diverse groups, thereby expanding the dataset and capturing a wider range of viewpoints. Third, imagining plants as long-term stakeholders proved difficult for participants, who viewed this scenario as distant and abstract. One way to address this gap is to develop phased, multi-level scenarios that guide participants from immediate, tangible care actions toward more ambitious, long-term visions in which public spaces are genuinely responsive to the needs of multiple species. In this study, the use of three future scenarios was particularly effective in unpacking the full spectrum of challenges identified, suggesting a promising direction for future iterations of the method. This limitation, as already articulated in previous chapters, is closely connected to the concept of more-than-human temporalities, which remain difficult to fully grasp and integrate.

5.6 Conclusions

This chapter has shown how the theoretical concepts developed throughout the book, particularly those articulated through the *Relational Atlas*, can be operationalised within a participatory design process situated in a specific context. The case study conducted in Nolo illustrates how knowledge of plant *stressors* and relational dynamics can be translated into participatory practices that foreground the needs and trajectories of nonhumans. In doing so, it shows how abstract ecological insights can become tangible components of community-based design work.

Although the approach remains oriented more toward designing *for* nonhumans than *with* them, the integration of speculative design, design fiction, and customised *provotypes* enabled participants to temporarily take into account as "representatives" nonhuman perspectives and reconsider the role of plants within urban environments. These methods provided a structured yet imaginative framework through which participants could explore alternative multispecies futures and challenge established assumptions about urban ecosystems.

The activity further demonstrated that situated communities are both willing and capable of engaging critically with multispecies concerns when supported by accessible and provocative tools, suggesting that participatory and speculative approaches hold substantial potential for fostering more relational and ecologically attuned forms of urban design.

References

1. Haraway D (1988) Situated knowledges: the science question in feminism and the privilege of partial perspective. Fem Stud 14:575–599
2. Manzini E (2015) Design, when everybody designs: an introduction to design for social innovation. MIT Press, Cambridge, MA
3. Florida R (2005) Cities and the creative class. Routledge
4. Meroni A (2007) Creative communities. People inventing sustainable ways of living. Poli. Design, Milan
5. Björgvinsson E, Ehn P, Hillgren P-A (2010) Participatory design and democratizing innovation. In: Proceedings of the 11th Biennial participatory design conference, pp 41–50
6. Akama Y, Light A, Kamihira T (2020) Expanding participation to design with more-than-human concerns, pp 1–11
7. Bastian M, Jones O, Moore N, Roe E (2016) Participatory research in more-than-human worlds. Taylor & Francis

8. Binder T, Brandt E, Ehn P, Halse J (2015) Democratic design experiments: between parliament and laboratory. CoDesign 11:152–165
9. DiSalvo C (2010) Design, Democracy and Agonistic Pluralism. In Durling D, Bousbaci R, Chen L, Gauthier P, Poldma T, Roworth-Stokes S, Stolterman E (eds) Design and Complexity - DRS International Conference 2010, 7-9 July, Montreal, Canada
10. Thorpe A, Rhodes S (2018) The public collaboration lab—infrastructuring redundancy with communities-in-place. She Ji J Des Econ Innov 4:60–74
11. White DF (2019) Ecological democracy, just transitions and a political ecology of design. Environ Values 28:31–53
12. Celermajer D, Schlosberg D, Rickards L et al (2021) Multispecies justice: theories, challenges, and a research agenda for environmental politics. Environ Polit 30:119–140
13. Fitz-Henry E (2022) Multi-species justice: a view from the rights of nature movement. Environ Polit 31:338–359
14. Botero A, Karasti H, Saad-Sulonen J et al (2019) Drawing together, Infrastructuring and politics for participatory design. University of Oulu, Oulu
15. Fuad-Luke A (2013) Design activism: beautiful strangeness for a sustainable world. Routledge
16. Hillgren P-A, Seravalli A, Eriksen MA (2016) Counter-hegemonic practices; dynamic interplay between agonism, commoning and strategic design. Strateg Des Res J 9:89–99
17. Markussen T (2013) The disruptive aesthetics of design activism: enacting design between art and politics. Des Issues 29:38–50
18. Mouffe C (2013) Agonistics: thinking the world politically. Verso Books
19. Munthe-Kaas P (2015) Agonism and co-design of urban spaces. Urban Res Pract 8:218–237. https://doi.org/10.1080/17535069.2015.1050207
20. Tassinari V, Staszowski E (2020) Designing in dark times: An Arendtian Lexicon. Bloomsbury Publishing
21. Escobar A (2018) Designs for the pluriverse: radical interdependence, autonomy, and the making of worlds. Duke University Press
22. Tsing AL, Deger J, Keleman Saxena A, Zhou F (2020) Feral atlas: the more-than-human Anthropocene. Stanford University Press
23. Stengers I (1997) Power and invention: situating science. University of Minnesota Press
24. Teli M, Foth M, Sciannamblo M, et al (2020) Tales of institutioning and commoning: participatory design processes with a strategic and tactical perspective. In: Proceedings of the 16th participatory design conference 2020-participation (s) otherwise, vol 1, pp 159–171
25. Huybrechts L, Benesch H, Geib J (2017) Institutioning: participatory design, co-design and the public realm. CoDesign 13:148–159. https://doi.org/10.1080/15710882.2017.1355006
26. Karasti H (2014) Infrastructuring in participatory design. In: Proceedings of the 13th participatory design conference: research papers, vol 1, pp 141–150
27. Huybrechts L, Devisch O, Tassinari V (2022) Beyond polarisation: reimagining communities through the imperfect act of ontologising. CoDesign 18:63–77
28. Forlano L (2016) Decentering the human in the design of collaborative cities. Des Issues 32:42–54
29. Monteiro M (2019) Ruined by design: how designers destroyed the world, and what we can do to fix it. Mule Design
30. Heitlinger S, Foth M, Clarke R, et al (2018) Avoiding ecocidal smart cities: participatory design for more-than-human futures, pp 1–3
31. Ehn P, Nilsson EM, Topgaard R (2014) Making futures. Marginal notes on innovation, design, and democracy. The MIT Press, Cambridge, MA
32. Ehn P (2008) Participation in design things. In: ACM digital library, pp 92–101
33. Gatto G, McCardle JR (2019) Multispecies design and ethnographic practice: following other-than-humans as a mode of exploring environmental issues. Sustainability 11:5032
34. Palmås K, von Busch O (2015) Quasi-Quisling: co-design and the assembly of collaborateurs. CoDesign 11:236–249

35. Rice L (2018) Nonhumans in participatory design. CoDesign 14:238–257
36. Tolbert D, Buzzanell PM, Zoltowski CB et al (2016) Giving and responding to feedback through visualisations in design critiques. CoDesign 12:26–38
37. Tomico O, Wakkary R, Andersen K (2023) Living-with and designing-with plants. Interactions 30:30–34
38. Veselova E, Gaziulusoy İ (2021) Bioinclusive collaborative and participatory design: a conceptual framework and a research agenda. Des Cult 14:149–183
39. Von Essen E, Allen MP (2017) Solidarity between human and nonhuman animals: representing animal voices in policy deliberations. Environ Commun 11:641–653
40. Roudavski S (2021) Interspecies design. In: Cambridge companion to literature and the Anthropocene, pp 147–162
41. Allen J, Weinrich M, Hoppitt W, Rendell L (2013) Network-based diffusion analysis reveals cultural transmission of lobtail feeding in humpback whales. Science 340:485–488
42. Ramsey G (2013) Culture in humans and other animals. Biol Philos 28:457–479
43. Burayidi MA, Allen A, Twigg J, Wamsler C (2020) The Routledge handbook of urban resilience. Routledge
44. Pitt H (2017) An apprenticeship in plant thinking. In: Participatory research in more-than-human worlds, pp 92–106
45. Frawley JK, Dyson LE (2014) Animal personas: acknowledging nonhuman stakeholders in designing for sustainable food systems, pp 21–30
46. Jørgensen IKH, Wirman H (2016) Multispecies methods, technologies for play. Digit Creativity 27:37–51
47. Metcalfe D (2015) Multispecies design (Doctoral dissertation). University of the Arts London and Falmouth University, London
48. Turner J, Morrison A (2020) Designing slow cities for more than human enrichment: dog tales—using narrative methods to understand co-performative place-making. Multimodal Technol Interact 5:1
49. Westerlaken M (2020) Imagining multispecies worlds. Malmö University
50. Westerlaken M, Gualeni S (2016) Becoming with: towards the inclusion of animals as participants in design processes. In: Mancini (ed) ACI '16: Proceedings of the Third International Conference on Animal-Computer Interaction, pp 1–10. Association for Computing Machinery, New York
51. Pitt H (2014) On showing and being shown plants. A guide to methods for more-than-human geography. Area 47:48–55
52. Rosińska M, Szydłowska A (2019) Zoepolis: non-anthropocentric design as an experiment in multi-species care. Nordes
53. Roudavski S (2020) Multispecies cohabitation and future design. In Boess S, Cheung M, Cain R (eds) Synergy - DRS International Conference 2020. https://doi.org/10.21606/drs.2020.402
54. Dunne A, Raby F (2013) Speculative everything: design, fiction, and social dreaming. The MIT Press, Cambridge, MA; London
55. Nijs G, Laki G, Houlstan R, et al (2020) Fostering more-than-human imaginaries: introducing DIY speculative fabulation in civic HCI, pp 1–12
56. Rosner DK (2018) Critical fabulations: reworking the methods and margins of design. MIT Press
57. Latour B (2018) Down to Earth: politics in the new climatic regime. Wiley
58. Wakkary R (2021) Things we could design: for more than human-centered worlds. MIT Press
59. Tanasescu M (2014) Rethinking representation: the challenge of nonhumans. Aust J Polit Sci 49:40–53
60. Tomitsch M, Fredericks J, Vo D et al (2021) Nonhuman personas: including nature in the participatory design of smart cities. Interact Des Architect 50
61. Hall M (2011) Plants as persons: a philosophical botany. Suny Press

62. Gibson P, Baylee B (2018) Covert plants: vegetal consciousness and agency in an anthropocentric world. Punctum Books
63. Gagliano M, Abramson CI, Depczynski M (2018) Plants learn and remember: lets get used to it. Oecologia 186:29–31
64. Gagliano M (2013) Persons as plants: ecopsychology and the return to the dream of nature. Landscapes 5:14
65. Nyers P (2007) Introduction: why citizenship studies. Citizsh Stud 11:1–4
66. Häkli J (2018) The subject of citizenship—Can there be a posthuman civil society? Polit Geogr 67:166–175
67. Thomas V, Remy C, Bates O (2017) The limits of HCD: reimagining the anthropocentricity of ISO 9241-210, pp 85–92
68. Fassi D, Vergani F (2022) Designing proximity with situated stakeholders. In: DRS2022: Bilbao, 25 June–3 July, Bilbao, Spain. Design Research Society, p 177
69. Fassi D, Vergani F (2020) Designing solutions for the commons. In: Sustainability awareness and green information technologies. Springer, pp 463–477
70. Manzini E (2019) Politics of the everyday. Bloomsburry
71. Carmona M (2015) Re-theorising contemporary public space: a new narrative and a new normative. J Urban 8:373–405
72. Jacobs J (1961) The death and life of great american cities. Random House, New York
73. Whyte WH (1980) The social life of small urban spaces. The Conservation Foundation
74. Deshpande P (2016) On potential application of interaction design for placemaking. In: Architecture and interaction. Springer, pp 321–343
75. Fassi D, Manzini E (2021) Project-based communities: lessons learned from collaborative city-making experiences. CoDesign 1–12
76. Manzini E, Rizzo F (2011) Small projects/large changes: participatory design as an open participated process. CoDesign 7:199–215
77. Fassi D, Sedini C (2017) Design actions with resilient local communities: goals, drivers and tools. Strateg Des Res J 10:36–46
78. Manzini E (2021) Abitare la prossimità. Idee per la città dei 15 minuti. EGEA
79. Mouffe C (2009) Democracy in a multipolar world. Millennium 37:549–561
80. DiSalvo C, Lukens J (2011) Nonanthropocentrism and the nonhuman in design: possibilities for designing new forms of engagement with and through technology. From social butterfly to engaged citizen: urban informatics, social media, ubiquitous computing, and mobile technology to support citizen engagement 421
81. Mouffe C (2000) The democratic paradox. Verso
82. de la Bellacasa MP (2017) Matters of care: speculative ethics in more than human worlds. University of Minnesota Press
83. Tassinari V, Vergani F (2023) Designing situated vocabularies to counter social polarizations: a case study of Nolo neighbourhood, Milan. Urban Planning 8
84. Boer L (2011) Participatory provocations. In: Proceedings of participatory innovation conference, pp 21–26
85. Mogensen P (1992) Towards a provotyping approach in systems development. Scand J Inf Syst 4:5
86. Sanders EB-N, Stappers PJ (2014) Probes, toolkits and prototypes: three approaches to making in codesigning. CoDesign 10:5–14
87. Sterling B (2013) Patently untrue: fleshy defibrillators and synchronised baseball are changing the future. Wired UK
88. Bleecker J (2022) Design fiction: a short essay on design, science, fact, and fiction. In: Machine learning and the city: applications in architecture and urban design, pp 561–578
89. Tonkinwise C (2015) Being dogmatic about defining speculative critical design future fiction. In: Medium. https://medium.com/@camerontw/just-design-b1f97cb3996f. Accessed 1 Nov 2021

90. Pólvora A, Nascimento S (2021) Foresight and design fictions meet at a policy lab: an experimentation approach in public sector innovation. Futures 128:102709
91. Carroll JM (1995) Scenario-based design: envisioning work and technology in system development. Wiley
92. Jégou F, Manzini E (2008) Collaborative services. Social innovation and design for sustainability. Poli.Design, Milan
93. Fassi D, Vergani F (2024) Design for public engagement: merging the role of academics and neighbourhood communities. In: Designing ethically in a complex world. Multiple challenges within design for public and social systems. FrancoAngeli, pp 85–99
94. Tassinari V, Piredda F, Vergani F (2025) Situated vocabularies. A situated and agonistic conversational platform. FrancoAngeli
95. Clément G (2004) Manifesto of the third landscape. Editions Sujet/Objet
96. Latour B (1993) We have never been modern. Harvard University Press
97. Voros J (2003) A generic foresight process framework. Foresight 5:10–21
98. Mancuso S (2021) The nation of plants. Other Press, LLC
99. Viola A (2020) Flower power. Le piante e i loro diritti. Passaggi Einaudi
100. Chapman CN, Milham RP (2006) The personas' new clothes: methodological and practical arguments against a popular method. In: Proceedings of the human factors and ergonomics society annual meeting. SAGE Publications, Los Angeles, CA, pp 634–636
101. Miaskiewicz T, Kozar KA (2011) Personas and user-centered design: how can personas benefit product design processes? Des Stud 32:417–430

Chapter 6
Designing in a More-Than-Human World: Reflections and Tips

Abstract This final chapter gathers and connects the main theoretical and empirical concepts that have emerged throughout the volume, intending to outline concluding reflections and suggestions for design from a more-than-human perspective. Across its various stages, the book has shown that recognising the agency of plants and other nonhumans does not mean romanticising the old idea of *nature* or attributing human capacities to other agents, but instead taking the complexity of multispecies relations as a starting point for rethinking design. The analyses presented, together with tools such as the *Relational Atlas* and the experiments conducted in urban contexts, have highlighted the need for design to be capable of negotiating, listening, slowing down, and attuning to the ecological interdependencies that shape our ways of inhabiting the world.

6.1 Introduction

More-than-human design is, at its core, an extension of the very essence of designing: a *relational act*. But why has this perspective become so urgent today? When we observe our existence on a planetary or even cosmic scale, our daily human actions appear almost negligible, exposed to forces that are greater than ourselves. At the same time, even within human–human relations, mutual understanding is necessarily partial; each individual inhabits a distinct *perceptual sphere* that mediates both what is received and what is expressed.

In this radical uncertainty, more-than-human design may appear to remain an inherently human-centred endeavour—and to some extent, it does. Nevertheless, two considerations render it necessary. First, ignoring the complex web of relationships that supports life on Earth puts human wellbeing at risk; when ecosystems collapse, the conditions that human societies depend on also fail. Second, broadening our perspective beyond our own species constitutes an act of recognition toward the myriad forms of life that support human existence daily, often without our

F. Vergani, *Reframing Urban Ecologies*, SpringerBriefs in Applied Sciences and Technology, https://doi.org/10.1007/978-3-032-18924-0_6

awareness. This chapter arises from this awareness: designing in a more-than-human, multispecies world means recognising the interdependencies that constitute us, assuming responsibility towards this bond, and imagining futures in which coexistence is not a constraint but a (re)generative possibility.

6.1.1 The Relational Space as a Domain of Design Negotiation

A first consideration in this volume is that designing, from a more-than-human perspective, is fundamentally framed as a *relational practice*. As discussed in Chap. 1, contemporary Design has long been moving away from a solutionist paradigm, centred on the production of immediate outcomes, towards a more processual, situated, and slower conception of practice [1, 2]. Within this framework, design concerns less the need to solve problems or create artefacts, and more the cultivation of relations among multiple life forms.

In this regard, recognising a systemic perspective among the various agencies that constitute a situated context (such as an urban public space like the one presented in Chap. 5) is of crucial importance. As previously extensively discussed, this relational process of care is constrained by both human and non-human perceptual biases, which significantly complicate its implementation. For these and many other limitations, more-than-human design, in its current understanding of designing *with* or designing *as*, still struggles to fully move beyond theory and take root in practice.

Moreover, including other species—or more broadly, other nonhuman agents—as participants in this relational space does not mean overlooking human concerns or idealising an ontological symmetry that does not exist. Instead, it means expanding the design space to incorporate forms of life that are typically excluded, in order to identify "matters of care" [3] that may guide new configurations of coexistence. Such an approach avoids the risk of slipping into ecological romanticism [4, 5], which can conceal paternalistic positions. It is therefore essential to acknowledge explicitly that more-than-human design remains, fundamentally, a human activity, ultimately directed towards human well-being. What shifts, however, is that this process simultaneously activates a practice that is attentive to multispecies interdependencies.

The micro-relational level (the everyday, situated, often invisible interactions) becomes the privileged ground on which to observe and rethink multispecies dynamics. Following Tsing [6], such interactions compose a world of *entanglements* in which no form of life exists in isolation. Recognising these ties does not entail imitating the other, but acknowledging the vital co-dependence [1] that binds all species, grounded in the fundamental drive for *survival*.

From this perspective, the *Relational Atlas* (RA) presented in the volume is also a theoretical framework that struggles to translate directly into practice. Its structure is informed by data and objectives that primarily address a human audience. Yet, it nonetheless incorporates nonhuman perspectives (that is, inevitably biased) based on existing knowledge of the needs and survival strategies of other species. Its value lies in its capacity to highlight relations as the genuine interspecies connectors, thereby informing the design of practices and tools that may subsequently enable processes of negotiation within real multispecies contexts.

6.1.2 Navigating Tensions: Western Perspectives and Indigenous Knowledges

Another theoretical thread running through this volume concerns the epistemological tensions between Western perspectives, which inform both the scientific literature and the design tools employed, and other worldviews such as Indigenous or locally rooted knowledges, which have long practised forms of multispecies relation without requiring academic legitimisation. As discussed in Chap. 1, the Western paradigm privileges the production of knowledge through quantification, empirical verification, and classification, constructing an idea of *nature* as an external object of study, separate from the human [7]. While this approach provides rigorous tools, it risks reducing the complexity of nonhuman agencies to interpretative models that favour simplification. Plant science—from classical plant physiology to more recent studies on plant neurobiology—is today highly advanced, yet still framed within human categories. The RA itself emerges from this compromise: interpreting *vegetal* actions through a human design language, while recognising that such translation inevitably transforms, simplifies, and at times even diminishes them.

By contrast, many Indigenous ontologies consider plants as agents endowed with intentionality, spirit, or personhood [8]. In Indigenous epistemologies, the world is not divided into nature and culture, but constituted by relationships in which humans and nonhumans participate in a shared fabric of life. This view resonates with Escobar's [1] notion of "radical interdependence," which invites us to understand the world as a mesh of co-constituting relations.

In attempting to integrate these perspectives, more-than-human and multispecies design occupies an interstitial position, seeking a balance between scientific rigour and ontological imagination. The challenge lies in developing design tools that refrain from reducing living beings to mere resources and instead maintain the capacity to recognise broader modes of interaction, including those not yet empirically verifiable and that continue to constitute significant gaps in current research. Within this context, the RA represents a step in this direction: it translates scientific knowledge into a relational language while leaving room for speculation, intuition, and a plurality of ways of knowing.

6.1.3 Different Spatial and Temporal Scales

A crucial reflection emerging from the more-than-human approach that permeates this work concerns the need to consider the multiple spatial and temporal scales through which multispecies relations unfold. More-than-human design operates across different scales, each characterised by distinct dynamics, temporalities, and possibilities for intervention. To grasp the full significance of this, these dynamics must be situated within a broader theoretical framework that integrates perspectives from ecology, anthropology, philosophy, and the environmental humanities.

As Fry [2] notes, designing with nonhumans raises important questions regarding the different rhythms of growth and evolution that characterise various agents. As discussed in Chap. 2, plants inhabit a different temporal scale, with significantly longer lifespans and much slower cycles of growth and reproduction. Drawing on Marder's [9] notion of *ontophytology*, which highlights the distinctive ways in which plants perceive and engage with time, it becomes clear that the temporal dimension of design must be reformulated when working within multispecies communities: rather than being understood as a process aimed at a predetermined goal, design (and, in particular, participatory design pathways such as the one described in Chap. 5) should be seen as an ongoing process that unfolds over time. A never-ending process not determined solely by humans but continuously shaped by the presence and actions of others.

Drawing on the concept of "ontological design" [10], we are encouraged to address temporality by considering the long-term effects of our actions on nonhumans and their environments, as well as their capacity to intervene in shaping physical space through temporalities that may be stretched or compressed. This, of course, raises important questions about how the temporalities of other species might be embedded within these processes. How can such unique modes of engagement be aligned? Is it even necessary to align them? And how vital is *attunement* to other temporalities?

The question of spatiality is likewise of central importance: as inherently nomadic beings, we often struggle to comprehend the extent to which agents, such as plants, are rooted in place and how profoundly their modes of spatial engagement differ from our own. Plant behaviours have evolved in response to the resources available in their immediate surroundings, whereas human behaviours have adapted to physically reaching resources through continuous movement. This, however, does not imply an absence of movement in plants, nor a diminished capacity to perceive or respond to what occurs around them. When spatiality is understood through the lens of relational space, a broader conception emerges. Relational space does not unfold solely at ground level (the layer primarily used by humans for interaction) but also encompasses what lies beneath (the soil) and what lies above (the air). Plants actively utilise these *strata* to establish and promote relationships with other life forms.

In contrast, humans engage with them in markedly different ways, particularly in the subsurface, which is typically treated as a service layer for concealing infrastructural systems, such as cables, pipes, and transport networks. Because soil is often insufficiently recognised as a fundamental substrate for the formation of interspecies interactions, human practices have constrained a wide range of plant activities. In the absence of adequate spatial conditions, plants are frequently compelled to grow into, and ultimately clash with, human infrastructures.

At a more abstract level, time and space also intersect with the notion of relational space, negotiation, and coexistence as defined within the RA. In this regard, the observational lenses function across multiple scales of perspective, enabling a layered understanding of multispecies interactions. The *micro-relational* scale represents the most immediate level at which multispecies coexistence manifests. Following Tsing [6], the everyday relations among plants, insects, animals, humans, and microorganisms constitute a dynamic and often invisible ecological fabric. It is here, through the *art of noticing*, that one observes proximity-based interactions: roots lifting pavements, leaves modulating microclimates, pollinators moving through the city, fungi and microorganisms maintaining soil fertility. This is the scale at which design can intervene with the most significant degree of accuracy. *Situated*, in the sense proposed by Haraway [11], it is also the scale most conducive to fostering relational care and to cultivating regenerative micro-ecologies. The RA is embedded at this level, translating plant behavioural patterns into strategies that can be interpreted and meaningfully engaged with by humans. Moving towards the *meso-scale*, complexity increases significantly. Here, we encounter urban ecological systems, including green networks, green corridors, public spaces, neighbourhood communities, and integrated infrastructures. As Hinchliffe and Whatmore [12] emphasise, the contemporary city is a more-than-human assemblage in which materials, living beings, technologies, and regulations interact in non-linear ways. At this scale, interactions no longer concern individual organisms but entire ecological and social communities. Designing at the *meso-scale,* therefore, requires engaging with political and economic tensions, complex regulatory frameworks, and often conflicting public perceptions.

At a *planetary scale*, viewed through an Earth-bound lens, multispecies relations are governed by ecological dynamics far beyond the possibility of direct control. Climate change, biodiversity loss, and pervasive pollution represent global forces that profoundly shape nonhuman life, including plants. Morton [13] describes such phenomena as *hyperobjects*: massive, diffuse, temporally extended, and difficult-to-localise entities that elude human and design agency. In this context, the contribution of design is not to solve these crises but to cultivate forms of habitability that prepare us for coexistence within an altered planet.

Finally, a last lens to consider, though virtually impossible to address operationally, is the *cosmic scale*, which situates life within temporalities that exceed not only design action but biological existence itself. This perspective foregrounds the fragility of the living and invites recognition of the contingency of terrestrial life within *Gaia*. While it offers no direct operational opportunities for design, the *cosmic scale* plays an epistemic role, repositioning the human in a state of relative insignificance in relation to geological and temporal forces that are vastly more extensive. Precisely because they are impossible to manage, these forces amplify the value of action at the micro-relational scale.

The specific contribution of more-than-human design, therefore, does not lie in its capacity to govern global dynamics, but in its attention to and cultivation of situated relations. These relations constitute the *entry points* through which plural and regenerative worlds become imaginable.

6.1.4 Translation, Transformation, and the Risk of Trivialising Nonhuman Agencies

As previously stated, embracing a nonhuman perspective involves a process of translation. It could be argued that a form of translation also takes place among humans, though to a reduced extent: whenever a message is conveyed or received, something is unavoidably lost or transformed. However, since nonhuman agencies are even more challenging to recognise because people cannot directly access them, they are often understood through simplified categories, analogical models, or speculations. As Gagliano et al. [14] argue, the potential for attributing perception or intentionality to plants poses a conceptual limitation that conventional language struggles to overcome. This is why practices of *noticing* and *attuning* are necessary in a more-than-human framework. They reactivate channels of communication in people that are now rarely, if ever, used.

The RA precisely positions itself along this trajectory by articulating a shared vocabulary (*actions*, *reactions*, *interactions*, and *strategies*) that enables humans to understand specific behavioural patterns of nonhumans, in this case, plants. While fully aware of the risk of simplification, the model makes visible forms of vegetal agency that would otherwise remain difficult to discern, translating emerging biological knowledge into a mediating device between theory and design practice.

Within this translational process, design has increasingly adopted the anthropomorphisation of nonhumans through human-centred tools that attribute personality traits, needs, and aspirations. Tools such as nonhuman personas [15, 16] are now used in design processes to incorporate nonhuman perspectives and ensure their coherence within the envisioned experience. Although these tools are frequently criticised for potentially reinforcing harmful anthropocentrism—a tendency historically amplified by design—they nonetheless hold the potential to make certain forms of nonhuman knowledge more explicit and accessible.

In this volume, Chap. 5 exemplifies this point: the use of a tool (the *legal form*) as an interface between human and nonhuman worlds proved effective in helping human stakeholders recognise these claims, prompting them to interpret the role of "representatives" and to engage with vegetal needs. These forms, essentially a more speculative reworking of nonhuman personas and situated within a narrative world amplified through design fiction, formalised intangible rights for the plants present in the context. This process not only exposed the absence of legal protection frameworks for nonhumans but also encouraged participants to engage critically with questions of multispecies justice.

6.1.5 Interspecies Clashes in the City

This volume has positioned plants at the core of its analysis as active agents within urban contexts, taking their agency as a foundation for constructing a relational framework with the potential to inform the development of new tools for more-than-human and multispecies design. Nonetheless, speaking of plants as agents may appear counterintuitive to many. This stems from the anthropocentric biases discussed in Chap. 2, such as the tendency to conceptualise *nature* and, particularly, plant life, through an *aestheticising* and romanticising lens [17]. Plants are frequently perceived as decorative and beneficial features for humans, obscuring their active role and preventing the recognition of potential tensions that may characterise cohabitation with other species, including our own. Such a misconception contributes to a superficial care for interspecies relations, limiting our ability to address the complexities of the urban environment in a systemic and informed manner.

Chapter 4 has shown how plant agency can generate stressors for humans, from roots lifting pavements, to pollen allergies, to maintenance costs and impacts on urban infrastructure. These examples reveal that articulating multispecies coexistence is far from a linear process; rather, it is a complex dynamic requiring negotiation, reciprocal adaptation, and attention to the diverse needs at play. At the same time, the environmental conditions produced by humans also constitute significant stressors for plants in cities, resulting in unbalanced and often problematic relationships.

When the analytical lens extends beyond the plant realm, the level of complexity increases substantially. Urban environments accommodate numerous additional forms of life with which humans frequently maintain far more conflictual relationships, including insects, rats, pigeons, as well as boars, wolves, and bears when they approach inhabited areas. In these cases, the cultural divide between "friendly species" (such as dogs, cats, butterflies) and "enemy species" becomes especially visible and profoundly shapes social, political, and design responses. Such examples illustrate that multispecies coexistence in urban space is never peaceful by default; it is the outcome of ongoing dynamics of competition, adaptation, and confrontation.

It is in this context that Peña-Guzmán's [18] concept of *diplomacy* becomes particularly relevant. Interspecies diplomacy calls for recognising animals and other forms of life as subjects with their own interests, requiring practices of negotiation and coexistence that go beyond logics of control or elimination. It is a relational ethics aimed at establishing non-hierarchical, cooperative relations grounded in respect, recognition, and active listening to the needs of other species.

Adopting the perspective of interspecies diplomacy, therefore, entails a radical rethinking of urban governance and design. It means questioning how to maintain conditions that reduce conflict, enhance ecological interdependencies, and support more equitable and sustainable forms of coexistence between human and nonhumans, with all the complexities such reasoning entails. How multispecies can an urban context truly be? Where is the boundary between the freedom of nonhuman action and the human tendency to impose order, cleanliness, and safety? And what value does human aesthetics hold when weighed against nonhuman functionality?

6.2 Final Consideration. The Plural Role of the Designer in a More-Than-Human Perspective

As illustrated in Chap. 5, the work conducted in *Piazzetta Transiti* highlighted how the role of the researcher and designer takes on multiple configurations within participatory processes oriented towards more-than-human design. While some functions were immediately recognisable, such as facilitating the process or designing tools, materials, and content, other dimensions emerged progressively during the collaboration with the local community. In this context, I also found myself acting as a communicator of knowledge on plant life, as an activist engaged in caring for the place, and as one of the founding members of the coalition of residents and organisations involved. This plurality of roles resonates with the figure of the designer as a combination of expert, facilitator, activist, and provocateur. In particular, the provocateur [19] is the one who brings current problems to light or makes them real so that they can figure out what needs to be done in the future. In line with Schön's [20] "reflective practitioner", the designer does not merely solve problems but helps them emerge, making them visible, debatable, and open to negotiation.

Alongside this transformative dimension, there emerges a more critical and attentive attitude characteristic of design approaches that extend beyond the human. From this perspective, the designer is not only an expert but must also assume the position of what Stengers [21] calls the *idiot*: a figure who slows down processes, suspends judgment, questions what appears self-evident, and listens to what ordinarily remains unheard. This posture resonates with Haraway's [4] call to "stay with the trouble", that is, to inhabit complexity consciously without seeking shortcuts or immediate solutions.

Moreover, Stengers' *idiot* recalls the designer as *careful*: someone who shows attention, care, and responsibility towards the human and nonhuman agents involved. In this sense, the figure of the designer is redefined as an *ecosocial mediator* who observes, notices, and acknowledges the radical interdependence among life forms. The integration of these two postures—the provocative and the *idiotic*—defines a designer who operates simultaneously as an agent of transformation and as a fellow guardian of complexity. On the one hand, designers stimulate new ways of thinking, generate discontinuities, and stage alternatives; on the other, they decelerate action to allow multispecies relations to surface and find space within the design process.

Based on these considerations, the role of the designer—and also that of the design researcher—may be rethought as that of an activator of critical reflection within situated and more-than-human communities. Through theoretical devices, visual tools, critical fabulation, speculative narratives, and participatory activities, designers can help make nonhuman agencies visible, tangible and alive, opening spaces for collective discussion around complex transformations, such as those involved in multispecies coexistence in urban space. In this sense, the work presented in this volume offers an example of how design can support processes of change not by merely producing solutions, but by cultivating the languages, sensibilities, and imaginaries necessary to radically rethink how we co-inhabit the world.

References

1. Escobar A (2018) Designs for the pluriverse: radical interdependence, autonomy, and the making of worlds. Duke University Press
2. Fry T (2009) Design futuring. University of New South Wales Press, Sydney, pp 71–77
3. de la Bellacasa MP (2017) Matters of care: speculative ethics in more than human worlds. University of Minnesota Press
4. Haraway D (2016) Staying with the trouble: making kin in the Chthulucene. Duke University Press
5. Morton T (2009) Ecology without nature: rethinking environmental aesthetics. Harvard University Press
6. Tsing AL (2015) The mushroom at the end of the world: on the possibility of life in capitalist ruins. Princeton University Press
7. Descola P (2013) The ecology of others. Prickly Paradigm Press
8. Harvey G (2014) The handbook of contemporary animism. Routledge
9. Marder M (2013) Plant-thinking: a philosophy of vegetal life. Columbia University Press
10. Willis A-M (2006) Ontological designing. Design Philos Pap 4:69–92
11. Haraway D (1988) Situated knowledges: the science question in feminism and the privilege of partial perspective. Fem Stud 14:575–599
12. Hinchliffe S, Whatmore S (2006) Living cities: towards a politics of conviviality. Sci Cult 15:123–138
13. Morton T (2013) Hyperobjects: philosophy and ecology after the end of the world. University of Minnesota Press
14. Gagliano M, Abramson CI, Depczynski M (2018) Plants learn and remember: lets get used to it. Oecologia 186:29–31
15. Lutz D (2023) The Nonhuman Persona Guide: how to create and use personas for nature and invisible humans to respect their needs during design. Damien Lutz
16. Tomitsch M, Fredericks J, Vo D et al (2021) Nonhuman personas: including nature in the participatory design of smart cities. Interact Des Architect 50:102–130
17. Morton T (2007) Ecology without nature: rethinking environmental aesthetics. Harvard University Press
18. Peña-Guzmán D (2021) Cetacean cultural rights: a third generation of rights at sea. Anim Law Rev 27:83
19. Mogensen P (1992) Towards a provotyping approach in systems development. Scand J Inf Syst 4:5
20. Schön DA (1983) The reflective practitioner: how professionals think in action. Basic Books
21. Stengers I (2005) The cosmopolitical proposal. In: Making things public: atmospheres of democracy, p 994

MIX
Papier aus verantwortungsvollen Quellen
Paper from responsible sources
FSC® C105338

If you have any concerns about our products,
you can contact us on
ProductSafety@springernature.com

In case Publisher is established outside the EU,
the EU authorized representative is:
Springer Nature Customer Service Center GmbH
Europaplatz 3, 69115 Heidelberg, Germany

Printed by Libri Plureos GmbH
in Hamburg, Germany